OPPENHEIMER
and the MANHATTAN PROJECT

Insights into
J. Robert Oppenheimer,
"Father of the
Atomic Bomb"

T0331613

OPPENHEIMER and the MANHATTAN PROJECT

Insights into
J. Robert Oppenheimer,
"Father of the
Atomic Bomb"

Editor
CYNTHIA C. KELLY
President, Atomic Heritage Foundation, USA

World Scientific

NEW JERSEY • LONDON • SINGAPORE • BEIJING • SHANGHAI • HONG KONG • TAIPEI • CHENNAI

Published by

World Scientific Publishing Co. Pte. Ltd.

5 Toh Tuck Link, Singapore 596224

USA office: 27 Warren Street, Suite 401-402, Hackensack, NJ 07601

UK office: 57 Shelton Street, Covent Garden, London WC2H 9HE

British Library Cataloguing-in-Publication Data
A catalogue record for this book is available from the British Library.

First published 2006
Reprinted 2008

OPPENHEIMER AND THE MANHATTAN PROJECT
Insights into J Robert Oppenheimer, "Father of the Atomic Bomb"

ISBN-13 978-981-256-418-4
ISBN-10 981-256-418-7
ISBN-13 978-981-256-599-0 (pbk)
ISBN-10 981-256-599-X (pbk)

Typeset by Stallion Press
Email: enquiries@stallionpress.com

Printed in Singapore

CONTENTS

Introduction

DEFYING THE ODDS

Cynthia C. Kelly
President of the Atomic Heritage Foundation

Investing two billion dollars in an attempt to build an atomic bomb in the midst of World War II was a serious gamble. While physicists understood that enormous energy would be released when the nucleus of an atom was split, harnessing that energy would be an immensely complex challenge. The odds of accomplishing this feat before the end of the war were slim.

Cynthia C. Kelly

When General Leslie Groves decided to choose J. Robert Oppenheimer to lead the project, most who knew Oppenheimer were skeptical. While Oppenheimer was widely acknowledged as a brilliant theoretical physicist, he had little management experience to prepare himself for the task of directing what would be the most ambitious scientific and engineering undertaking of the twentieth century.

At Berkeley, Oppenheimer had a reputation for being a "smart aleck," often arrogant and impatient with those who could not keep up with him intellectually. Could Oppenheimer recruit and lead a team of hundreds of scientists and engineers, military and support staff under highly pressured and trying circumstances? These

Oppenheimer at Berkeley

reservations, coupled with Oppenheimer's alleged Communist Party affiliations, made his successful tenure as director of the project's laboratory at Los Alamos seem improbable.

Despite the long odds, the atomic bomb was produced in time to bring an end to the war. Oppenheimer quickly proved himself to be an exquisite manager of people, able to effectively motivate and deploy even those with problematic personalities. While Robert R. Wilson expressed grave doubts at first, "He had style and he had class," Wilson told an interviewer in 1982, "He was a very clever man. And whatever we felt about his deficiencies, in a few months, he had corrected those deficiencies."[1] Oppenheimer's charisma, brilliance, and personal command of all aspects of the project made him an extraordinary leader.

The government's two-billion-dollar gamble paid off, bringing an end to World War II and establishing the United States as a super power. However, preserving some of the tangible properties of the Manhattan Project continues to face long odds. Across the nation, most of the last remaining properties from the Manhattan Project owned by the Department of Energy are slated to be demolished as part of the environmental cleanup of the nuclear weapons complex.

S Site, Room 24
Courtesy of the Los Alamos Historical Society

At Los Alamos, the Laboratory has identified a dozen properties of the fifty properties left in its study called "Sentinels of the Atomic Dawn." While $700,000 is available from a Save America's Treasures grant, additional funds will be needed to stabilize, protect and restore these properties for future generations.

At Oak Ridge, the K-25 plant was the world's first gaseous diffusion plant and produced enriched uranium for the Hiroshima bomb. A U-shaped building with each of its arms extending one-half mile, the K-25 plant and over one hundred other properties at the site are slated for demolition. Discussions

---

[1] Bird and Sherwin, quote from Robert R. Wilson interview with Owen Gingrich, 23 April 1982, p. 4.

between the Department of Energy, the Advisory Council for Historic Preservation, the Commonwealth of Tennessee and concerned parties are currently focused on preserving the North End of the K-25 plant, three of the 54 units that make up the plant. However, funds need to be found very soon to preserve this section or in a few years there will be nothing left of one of the engineering marvels of the twentieth century.

At Hanford, the fate of the B Reactor along the shores of the Columbia is scheduled to be decided by the fall of 2005. Itself an engineering marvel, the B Reactor was designed by Enrico Fermi and his team of physicists at the University of Chicago. Pressured by President Franklin Roosevelt to take on the task of constructing the reactor, the DuPont Company then translated these concepts into engineering blueprints and assembled 50,000 workers to build and operate it.

While the B Reactor has been open to the public for nearly twenty years, the Department of Energy is planning to "cocoon" it, a process that would destroy the reactor building and its historic integrity. The only prospect of preserving it is to find an organization willing and able to commit to its long-term operation and maintenance. One option is to incorporate the B Reactor and other properties from the Manhattan Project into the national park system.

To investigate this alternative, Senators Domenici and Bingaman have sponsored legislation to authorize a study to explore the feasibility of creating a national park unit at one or more of the Manhattan Project sites.[2] The study will explore various management alternatives with continuing roles for the Department of Energy as well as other Federal, State and local agencies that have or may want to play various roles at these sites.

In addition, the FY 2004 Energy and Water Appropriations Act provided one million dollars to take urgent actions to preserve the Manhattan Project properties. The funds were intended to enable the Atomic Heritage Foundation to continue its efforts nationwide and for the Manhattan Project communities to address priorities such as capturing oral histories and stabilizing or restoring properties and artifacts that are in danger of being lost to future generations.

In August 2004, the Atomic Heritage Foundation completed its report to the Department of Energy on how best to preserve the most significant

---

[2]The Manhattan Project National Historic Park Site Study Act (PL 108-370).

Manhattan Project properties and this history.[3] The report considers the costs of restoration and long-term stewardship as well as alternative management strategies using local, state and other federal agencies, nonprofit organizations and private resources. Because a number of decisions are still pending and the difficulty of getting cost estimates for preservation, the report remains a "work in progress."

How long are the odds that we will be able to save some of the heritage of the Manhattan Project? We have made a lot of progress in the last five years but time is very short as the Department of Energy has many of these properties in its sights for demolition. Is there sufficient public support? On the one hand, there is enormous national interest in World War II. However, because of the cloak of secrecy concerning nuclear weapons production, many people do not know about the Manhattan Project and its role in World War II. Surveys of those who visit the immensely popular Spy Museum in Washington, DC, find that ninety percent of visitors do not know what the Manhattan Project was.

And yet, the development of the atomic bomb was one of the most significant events of the twentieth century. As Richard Rhodes commented at the Atomic Heritage Foundation's symposium in April 2002: "The closing days of the Second World War mark a turning point in human history, the point of entry into a new era when humankind for the first time acquired the means of its own destruction." The legacy of the atomic bomb has permeated every aspect of our lives and dominates world politics from Iraq to North Korea.

To understand the twenty-first century world we live in, we must understand the history of the atomic bomb and its indelible legacy. Having some of the tangible remains of the Manhattan Project will help to bring the public back to World War II when Hitler had taken over much of Europe and was thought to be developing an atomic bomb. Here, scientists and engineers had to draw upon their ingenuity, resourcefulness and determination as there were no high-speed computers or sophisticated electronics. They were truly working on the frontiers of science, with just microscopic quantities of uranium-235 and plutonium that needed to be analyzed and produced in quantities sufficient to fuel an atomic bomb.

---

[3] This report, "Preserving the Remains of the Manhattan Project," is available at http://www.atomicheritage.org/articles.htm or by calling the Atomic Heritage Foundation, 202-293-0045.

By witnessing first-hand the humble wooden sheds at Los Alamos where key components of the bomb were assembled, the public can begin to grasp some of the makeshift aspects of the effort. Preservation of some of the key properties is essential to understanding how the United States and her allies won the race to develop the world's first atomic bomb, changing the course of history forever.

The odds may be long but with the support and leadership of Senators Domenici and Bingaman, Governor Bill Richardson, the Los Alamos National Laboratory, Los Alamos County Council, Los Alamos Historical Society, and many other organizations that are working in partnership with the Atomic Heritage Foundation, I believe we can beat those odds. Like the Manhattan Project workers, we need to draw upon our ingenuity, resourcefulness and determination and work as a team.

Thanks for your interest and help in this venture.

CHAPTER ONE

Introducing Oppenheimer

OPPENHEIMER RECONSIDERED

The Honorable Jeff Bingaman
United States Senator from New Mexico

The story of Robert Oppenheimer is as timely as today's news and as timeless as a Greek tragedy. He was a brilliant scientist who devoted his talents to the service of his country. He was celebrated for making the atomic bomb and vilified for not wanting to make the hydrogen bomb. He helped unlock the secrets of the atom for his country and, in the end, his Government would not trust him with those secrets.

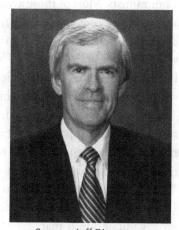

Senator Jeff Bingaman

His contributions to the Manhattan Project and to Los Alamos are legendary. He came up with the idea of a central weapons laboratory, and he picked the site for it, here at Los Alamos. Although there were many brilliant scientists and engineers who made enormous contributions to the Manhattan Project, Oppenheimer's contribution was unique. He was the Laboratory's first director; he recruited its original staff, and he led it to its wartime success.

Shortly after the war, Dr. Oppenheimer spoke eloquently of the Manhattan Project as having "led us up those last few steps to the mountain pass; and beyond there is a different country." He left Los Alamos and the Manhattan Project once the height was scaled, but he continued to help us find our way through the new country. He felt a deep responsibility for his work on the Manhattan Project and thought it was his duty to continue to make his technical experience and judgment available to the Government.

11

For nine years after the war ended, the Government drew heavily upon his talents. He served faithfully on numerous defense and nuclear policy committees. He chaired the General Advisory Committee to the Atomic Energy Commission. Under his leadership, the General Advisory Committee promoted the development of this Laboratory, the production and perfection of atomic weapons, and the development of nuclear reactors for submarines and naval propulsion. But he (and a majority of the General Advisory Committee) opposed the development of the hydrogen bomb.

It was Dr. Oppenheimer's opposition to the H-bomb, more than anything else, that made his opponents into enemies and fueled their suspicions of his loyalty. Undoubtedly, Oppenheimer had friends and relatives who were Communists. Most of those associations had been formed long before the war and most had long since ended. All of them had been thoroughly scrutinized by the Army when it cleared him in 1943 and by the Atomic Energy Commission when it cleared him in 1947. They now became the basis of new allegations. In December 1953, the Atomic Energy Commission formally charged him with disloyalty and suspended his security clearance.

Dr. Oppenheimer replied, with great dignity, that he had no desire to retain an advisory position if his advice was not needed, but that he could not ignore the suggestion that he was "unfit for public service." He decided to answer the charges against him and asked for a hearing to clear his name. What he got was not the objective "inquiry" called for by the Atomic Energy Commission's rules. It was a trial — there is no other word for it — and a grossly unfair one at that.

J. Robert Oppenheimer

The charges against Dr. Oppenheimer were long and complex. Most involved his past associations, which had already been had already been thoroughly and repeatedly invested. But the Commission went further and charged him with having "expressed" views opposing the development of the H-bomb. That was the crux of the matter.

Dr. Oppenheimer was tried, in secret, before a specially-appointed three-member personnel security board. He was prosecuted by an aggressive former criminal prosecutor specially retained for the case. The FBI bugged Oppenheimer's conversations with his lawyers and potential witnesses, and reported what it heard to the Commission. Evidence was withheld from Oppenheimer and his attorneys. Legal standards were lowered to meet the evidence. The whole affair was carefully orchestrated by the AEC's chairman, Lewis Strauss.

In the end, all three board members found Oppenheimer loyal, but two of the three concluded that he was a security risk and recommended that his security clearance not be restored. They found that his failure to give "enthusiastic support" to the H-bomb program and his "highly persuasive influence" among fellow scientists were not in "the strongest offensive military interests of the country."

Dr. Oppenheimer appealed the board's decision to the five-member Atomic Energy Commission. The Commission, by a four-to-one vote, found Oppenheimer to be loyal, but by a different four-to-one vote, found him to be a security risk. The Commission steered clear of the H-bomb charges, though they probably played a role in its decision. Instead, the majority based its decision on Oppenheimer's character and his associations.

On June 29, 1954, fifty years ago on Tuesday, the Atomic Energy Commission formally revoked Dr. Oppenheimer's clearance, forever ending his involvement in the atomic energy program. Ironically, Dr. Oppenheimer's term on the General Advisory Committee had expired two years before. His only remaining contact with the AEC was a consulting contract, which was scheduled to expire, along with his security clearance, the next day anyway.

History will be a fairer judge and will reach a truer verdict than the Commission. Robert Oppenheimer will be remembered, I believe, as a brilliant scientist who applied his talents loyally and unstintingly to our national defense. He will be remembered, too, as one who thought deeply about the forces unleashed by the Manhattan Project, and realized how essential it is for mankind to use wisely, in his words, "the new powers, the new alternatives, of an advancing mastery of nature" for "his welfare and his freedom, and not his destruction."

The clouds over Robert Oppenheimer's reputation have long since begun to dissipate. His many friends and supporters, both in the Government and in the scientific community, never doubted his loyalty. One such supporter was Senator Clinton P. Anderson. When President Eisenhower nominated

Dr. Oppenheimer's nemesis, Lewis Strauss, to be the Secretary of Commerce, Senator Anderson led the opposition to the nomination. Lewis Strauss had given Senator Anderson many reasons to oppose his nomination over the years, but his abusive treatment of Dr. Oppenheimer was chief among them. The Senate rarely rejects a cabinet nomination, but at Senator Anderson's urging, the Senate rejected Lewis Strauss' nomination in 1959.

In 1963, President Kennedy selected Dr. Oppenheimer to receive the Enrico Fermi award, which President Johnson bestowed on him after President Kennedy was assassinated. In 1994, the FBI publicly announced that allegations that Dr. Oppenheimer had shared secrets with the Soviets were "unfounded."

I have sought to add to these efforts by sponsoring, along with Senator Domenici and Senator Feinstein, a Senate resolution recognizing Dr. Oppenheimer's loyal service and contributions to the nation. The Senate unanimously agreed to the resolution Thursday evening.

In closing, I commend the Atomic Heritage Foundation for holding this conference and for its efforts to preserve the Manhattan Project properties here at Los Alamos and at other sites. I support these efforts and have sponsored legislation in the Senate to have the Secretary of the Interior consider adding the major Manhattan Project sites to the National Park System. The Senate Committee on Energy and Natural Resources approved the bill in April and it is now awaiting action by the full Senate. I think it is important that we save this significant part of our history and our heritage for future generations.

ROBERT OPPENHEIMER: KING OF THE HILL

Richard Rhodes
Author of "The Making of the Atomic Bomb" and "Dark Sun"

The Manhattan Project is fading into myth. Sad to say, the last of its first-rank leaders, Hans Bethe, today lies mortally ill. The letter from Einstein to Roosevelt eclipses the British MAUD Report. Los Alamos, a laboratory on a mesa surrounded by a wilderness, a small coterie of scientists witching historic transmutations, eclipses armies of workers and vast factories at Hanford and Oak Ridge. Hiroshima eclipses Nagasaki, poor Nagasaki, even as the war in Europe with its epic D-Day extravaganza eclipses the longer and crueler Pacific War. And to our point here

Richard Rhodes
Photo by Gail Evenari

today, Robert Oppenheimer, a century after his birth on April 2nd, 1904, is rapidly eclipsing General Groves and half a hundred others as the shining talent, the indispensable leader of the project, the Prospero of this historic Tempest.

The true history, as we all know, was far otherwise: The MAUD Report and three successive National Academy of Sciences reports turned the tide; the first bombs were designed and built at Los Alamos, to be sure, but the armies of workers and the vast factories produced their rare materials. Nagasaki suffered equally with Hiroshima for the Japanese leadership's refusal to surrender. Russian determination, Allied Lend-Lease and invasion achieved victory in Europe, but it needed atomic bombs to end the Pacific War. And no one who was part of the Manhattan Project, even within the close, intense community here on the Hill, doubted that General Groves was in charge. Nor did the project lack for other colorful characters, larger than life-sized: Bethe, Edward Teller, Ernest Lawrence, Enrico Fermi, Vannevar Bush, Arthur Compton, Leo Szilard, Harold Urey, Luis Alvarez, Emilio Segre, Eugene Wigner, Crawford Greenewalt, Paul Tibbets, Ken Bainbridge, I. I. Rabi, George Kistiakowsky, Deke Parsons, and of course Klaus Fuchs and many others — people whom I and many of you here knew in person, though I was not fortunate enough to meet Oppenheimer while he was alive.

It's worth asking why one man, Robert Oppenheimer, should emerge from such a rich and crowded field of candidates as the iconic central figure of what is arguably the single most important historic development of the twentieth century. I hope today's symposium of Oppenheimer experts will at least begin to answer that question, if an answer is possible to so obscure a phenomenon as the making of myth.

That Robert Oppenheimer's complexities should be reduced to a myth-ical unity is ironic; his contemporaries found him various indeed. Tall, thin, handsome, brilliant, with piercing blue eyes, chain-smoking, intense, elegant, witty, cruelly dismissive when he chose to be, generous, passionate, idealistic, but also divided within himself and by his own admission self-loathing, he seemed different men to different people. Edward Teller told me that Robert Oppenheimer was the finest lab director he had ever known, and I took that assessment seriously: praise from a man's worst enemy is praise indeed.

Hans Bethe told me Oppenheimer was able to direct the effort at Los Alamos so suc-cessfully because he was so much smarter than everyone else, and Bethe included himself in that comparison. Bethe told me also — a more telling insight, I think — that Oppenheimer had been casually cruel to people who made mistakes around him, including Bethe, before the war and after the war, but that he suspended the hostili-ties at Los Alamos.

Chester Barnard, president of New Jersey Bell, described Oppenheimer in

J. Robert Oppenheimer

1947 as "an extraordinary man who worked very hard and always seemed to be on the verge of a nervous breakdown."[4] His students and his friends saw him differently from his enemies, of course; to Lewis Strauss, Boris Pash and William Borden, among others, Oppenheimer was a Machiavellian schemer and a Communist spy. To Oppenheimer's enemies, in the terrible security hearing they imposed on him that condemned him to internal exile and destroyed him, tough-minded I. I. Rabi had the irrefutable rebuttal:

> The suspension of the clearance of Dr. Oppenheimer, [Rabi told the Gray Board,] was a very unfortunate thing and should not have been done [. . .] against a man who had accomplished what Dr. Oppenheimer has accomplished. There is a real positive record, the way I expressed it to a friend of mine. We have an A-bomb and a whole series of it, and what more do you want, mermaids?[5]

Rabi knew him well:

> He was an aesthete, [Rabi described Oppenheimer to Bill Moyers.] I don't think he was a security risk. I do think he walked along the edge of a precipice. He didn't pay enough attention to the outward symbols. He was a very American person of a certain kind. A certain kind of intellectual, aesthetic person of the upper middle classes.[6]

On another occasion Rabi assessed his friend's personal conflicts and their consequences for his science:

> I found him excellent, [Rabi said.] We got along very well. . . I enjoyed the things about him that some people disliked. It's true that you carried on a charade with him. He lived a charade, and you went along with it. It was fine — matching wits and so on. Oppenheimer was great fun, [Rabi goes on,] and I took him for what he was. I understood his problem. . . [His problem was] identity. . . He reminded me very much of a boyhood friend about whom someone said that he couldn't make up his mind whether to be president of the B'nai B'rith or the Knights of Columbus. Perhaps he really wanted to be both, simultaneously. Oppenheimer wanted every experience. In that sense, he never focused. My own feeling, [Rabi concludes,] is that if he had studied the Talmud and Hebrew, rather than Sanskrit, he would have

---

[4] Quoted in Richard Rhodes, *Dark Sun* (Simon & Schuster, 1995), p. 203.
[5] Quoted in *ibid.*, pp. 558–559.
[6] Quoted in *ibid.*, p. 559.

been a much greater physicist. I never ran into anyone who was brighter than he was. But to be more original and profound I think you have to be more focused.[7]

I am certainly not competent to judge if Oppenheimer's science was less original and profound than it might have been. Perhaps Rabi was. Others today will discuss other periods of Oppenheimer's life and career and perhaps address Rabi's contention as well. I want to look briefly at what I believe to be Oppenheimer's greatest achievement after the bomb itself, an original and profound achievement indeed, and neglected in something of the same way that the original discovery of the antibiotic properties of penicillin was neglected, resting in the files waiting to be pulled out and understood for what it is; the only ultimate answer to the hard, cruel fact of the bomb, to Curtis LeMay's "the bombers always get through" and William Borden's "there will be no time": I mean the Acheson–Lilienthal Report that Oppenheimer in 1946 guided his four colleagues on Acheson's panel of expert consultants to prepare.[8]

In the curious way of government reports — John Manley once commented sardonically that "quite contrary to the way I thought things were [in Washington,] you don't do staff work and then make a decision. You make a decision and then do the staff work" — the Acheson–Lilienthal Report originated in the Truman Administration's late-1945 Agreed Declaration with Britain and Canada to "prevent the use of atomic energy for destructive purposes" and to "promote the [. . .] utilization of atomic energy for peaceful and humanitarian ends." Such action required a plan; Jimmy Byrnes as Secretary of State got the job of devising it. Byrnes appointed a protesting Dean Acheson chairman of a committee that included General Groves, Vannevar Bush, James Bryant Conant and John J. McCloy. Acheson in turn appointed the five-man panel. David Lilienthal, chairman of the TVA, chaired it, and besides Oppenheimer, it included Monsanto chemist Charles Thomas, General Electric engineer Harry Winne and Chester Barnard — not exactly a bunch of wild-eyed internationalists.

---

[7] Quoted in *ibid.*, pp. 240–241.
[8] The discussion that follows is paraphrased from *ibid.*, pp. 229–233, where citations may be found for quotations.

Oppenheimer came prepared. He had explored the complexities of international control not only with Niels Bohr at Los Alamos and with Conant but also with Rabi. "Oppenheimer and I met frequently and discussed these questions thoroughly," Rabi said later. "Once [Oppenheimer] got interested in something, he went right on to become the leader of it." Gordon Arneson, the State Department's specialist on atomic matters, says Oppenheimer became "the chief teacher for the Acheson–Lilienthal group."

The men met first in Washington. Oppenheimer gave them a ten-day course in nuclear physics, properly taking control, as the only real expert, of defining the technical basis of the problem, but other than serving as their *savant* he kept his own council at first. They moved next to New York to talk to a group of scientists, including Luis Alvarez, who had explored for Groves a scheme of control by inspection alone, involving what we would now call *national technical means.*

Discussion intensified. Ideas came from every side — these were men of diverse background and conviction — and they debated them night and day. When patience gave way to exasperation and someone proposed simply outlawing the bomb, which happened frequently, Lilienthal always waved a newspaper clipping about the Agreed Declaration to remind them that their government had already committed itself to international control. Back to Washington to study geology. They made progress. Then they got seriously stuck. Lilienthal proposed they tour Oak Ridge and Los Alamos. Whiskey on the train down to Knoxville and a hung-over tour of the vast gaseous-diffusion plant, where supervisors prowled among the surrealistic piping on bicycles, warmed their friendship.

They flew to Los Alamos in Groves's private C-54. The President was trying to reach Lilienthal — to offer him a Cabinet post as Secretary of the Interior, the TVA chairman wrongly thought — but not even that provocation dulled him to the significance of the secret mesa, as he wrote,

General Leslie R. Groves and
David Lilienthal

. . .with the high mountains forming a majestic backdrop [where they] went into casual little buildings, saw things only few men have seen, talked with soft-spoken, gentle, intelligent men about the things they had done. . . Now I have sense, [Lilienthal concluded,] that this thing of atomic bombs is *real*. . .

Herb Marks, Acheson's personal representative to the panel, who accompanied the men on their travels, caught the mood and a whiff of the essence of the problem:

It wasn't a large place, [he wrote of the building here on the Hill where they examined the unassembled components of the few bombs yet in the stockpile,] . . . and it wasn't a spectacular one. I looked around me and there were the same materials, colors, textures and fabrics you might see in any warehouse. I saw the receptacles that contained the labor of God-knows-how-many men, the cargoes of thousands of freight cars, the mental triumphs of gifted scientists born in a dozen countries. The receptacles were small, and I thought to myself: *Hell!* I could walk out of here with one of them in my pocket. Not that I could have. Too many soldiers outside and inside the vault were watching us closely — tough troops who looked as though they kept their rifles cleaned. And supposing I had got away with one, what could I, an ordinary layman, have done with it? In a way, the same was true of so much of the whole Manhattan District.

Marks concludes:

It bore no relation to the industrial or social life of the country; it was a separate state, with its own airplanes and its own factories and its thousands of secrets. It had a peculiar sovereignty, one that could bring about the end, peacefully or violently, of all other sovereignties.

What that panel of hard-eyed, practical men came to was a radical proposal. Remarkably, it won their common agreement. When Bohr read it he wrote Oppenheimer of his "deep pleasure." In every word of it, he said, he found

just the spirit which I think offers the best hopes for the development in which we all put our whole faith.

Most of you know what was in that report, though I wonder every time I reread it if anyone in authority ever quite grasped the full importance of its argument. It was nothing less than an argument for abolition when no more than a few bombs had yet been built, anticipating the nuclear arms race and all its near misses and understanding that the bombers (or the missiles) always get through. "Any system," it argued, "based on outlawing the purely military development of atomic energy and relying solely on inspection for enforcement would at the outset be surrounded by conditions which would destroy the system." To the contrary, "every stage in the activity, leading from raw materials to weapons, needs some sort of control." If, for example, a putative international Atomic Development Authority were the only entity that could legally own and process uranium ore, then "not the purpose of those who mine or possess uranium ore but the mere fact of their mining or possessing it becomes illegal, and national violation is an unambiguous danger signal of warlike purposes. The very opening of a mine by anyone other than the international agency is a 'red light' without more; it is not necessary to wait for evidence that the product of that mine is going to be misused." And if the Authority spread its mines and factories and laboratories and reactors around the world, so that their benefits could be shared, then "the real protection will lie in the fact that if any nation seizes the plants or the stockpiles that are situated in its territory, other nations will have similar facilities and materials situated within their own borders so that the act of seizure need not place them at a disadvantage."

This remarkable idea — spreading the intrinsically dangerous mines and factories around — is indistinguishable from what has come to be called *nuclear proliferation*, except that the agent of proliferation in the Acheson–Lilienthal Report would have been an organ of the United Nations rather than individual states, and the technology that proliferated would have been infrastructure alone rather than infrastructure and stockpiled weapons. Though the report does not belabor the point, it notes more than once that true security is incompatible with secrecy. Its proposal for a radical system of self-policing makes starkly clear what the condition that Niels Bohr had called an "open world" would be: a world where how to design atomic bombs might be public knowledge (as it has come to some extent to be); a world, as it were, where the guns have all been laid out together in the open on a table, but disassembled, and arranged so as to be within everyone's equal reach. Would it have worked? In a much more unstable and dangerous form, as state-sponsored nuclear proliferation, it did, and does. It would work without warheads and weapons in a world with much greater transparency than presently obtains. We are moving rapidly toward that transparency as new technologies penetrate privacy and sweep secrets away; it's no great prediction to say that by 2050 the only privacy that will be left in the world will be the privacy we legislate for ourselves.

Oppenheimer and his colleagues' farsighted proposal may have been premature. Certainly the Soviet Union had no intention of abrogating atomic arms until it knew how to build them. Bernard Baruch, who modified the Acheson–Lilienthal Report into his Baruch Plan to present to the United Nations, missed the point completely when he complained that the Report "did not deal with the problem of enforcement." That's what the Report's "red light" was about: a move to commandeer any part of the nuclear infrastructure would be an act of war, to which other nations might respond accordingly. Baruch added a clause promising "swift and sure enforcement" and demanded that the Security Council give up veto authority in atomic matters, guaranteeing the rejection of his version of the plan. Oppenheimer, observing that the United Nations was hardly the place to forge agreement on the highest matters of national security policy, judged that his government was not serious about international control.

Oppenheimer chaired one more panel on disarmament, for Dean Acheson, in 1952. Vannevar Bush and Allen Dulles sat among its members; McGeorge Bundy served as secretary and rapporteur — once again, experienced and sober men, for whom only a paranoid could imagine that Oppenheimer might serve as a Svengali. As had the Acheson–Lilienthal Report, Oppenheimer's panel also came to a conclusion about the new knowledge of how to release nuclear energy that is as valid today as it was then, and as inescapably final:

> Fundamentally, and in the long run, the problem which is posed by the release of atomic energy is a problem of the ability of the human race to govern itself without war. There is no permanent method of excising atomic energy from our affairs, now that men know how it can be released. Even if some reasonably complete international control of atomic energy should be established, knowledge would persist, and it is hard to see how there could be any major war in which one side or another would not eventually make and use atomic bombs. In this respect the problem of armaments was permanently and drastically altered in 1945.[9]

Nuclear weapons made supposedly more useable and therefore credibly deterrent with lower yields will not solve the problem, nor will bunker busters,

---

[9] Quoted in *ibid.*, p. 588.

nor will missile defenses, nor will preemptive wars. Only the deterrence of knowledge in an open world — which means, practically speaking, nothing more utopian than delivery times from mothballed factory to target of three months rather than delivery times from silo to target of thirty minutes — only the deterrence of knowledge without stockpiles will buy the world the space it needs to come to its senses or act in concert whenever an entity bent on domination attempts to violate the ban. Which is another way of saying that the problem will never go away. Of course it won't: knowledge of how to release nuclear energy is new knowledge of the natural world, to which the human world has no choice but to adapt or be destroyed, just as knowledge of global warming is new knowledge of the natural world, just as knowledge of HIV and other scourges is new knowledge of the natural world. With Bohr, Oppenheimer understood that truth, and it was a deeper understanding than theoretical physics, original and profound.

Well. That's part of my understanding of who Robert Oppenheimer was. Speakers yet to come will have other views and other insights. I got to know Oppenheimer's protégé and close friend Robert (Bob) Serber late in Bob's life when we worked together editing *The Los Alamos Primer* for publication — they were Bob's lectures, and he thought they ought to be in print. After many meetings with him, when I thought I knew him a little, I finally worked up the courage to ask him if my portrait of his friend in my book *The Making of the Atomic Bomb* was at all accurate. Bob was a kind man. He thought for awhile, and then he said, "Well, of the books I've read about Oppie (this was several years ago) I'd say it's the least inaccurate."

So I'm looking forward to hearing the even less inaccurate portraits yet to come of a complex, charismatic man.

Thank you.

A NOVEL IDEA OF OPPENHEIMER

Joseph Kanon
Author of "Los Alamos" and "The Prodigal Spy"

Joseph Kanon

It's an honor for me to be with you today, and something of a surprise. I am not a historian, and certainly not a scientist. I may, in fact, be the only person here who had trouble with high school physics — basic high school physics. But life, as any physicist can tell you, is unpredictable. After I published my novel *Los Alamos*, which is what Graham Greene used to call an "entertainment," I found myself, with no scientific knowledge at all, being welcomed, taken in — a sort of second cousin, three or four times removed — to this extraordinary scientific community.

Many of you have written to me, or come up to me at book signings, and have shared anecdotes, stories about the Hill; sometimes as if I had actually been here with you on the Project. A few of the letters were scolding. It may be that because we now so readily accept spin in our public lives, we've come to demand more accuracy in our fiction — but in any case I'm not allowed to get away with much. I've been corrected about the location of the laundry in the Sundt units (this to prove that you couldn't hear people making love through the walls). I've been told the back road to Jemez wasn't paved in 1945 and that a roadhouse outside Nageezi — wholly imaginary, by the way — was in the wrong location. I've treasured all these letters, partly because I do like to get things right and partly because they suggest that the Los Alamos I imagined was reasonably close to the one you knew. But how do you get a person right? How do you get Oppenheimer right?

24

When the organizers of this symposium asked me to speak, I said, "You know, I didn't know Robert Oppenheimer" And they said, "But you knew him in a way no one else did — you had to make him up. Tell them how you did it." That sounded to me suspiciously like "How does one write fiction?", a subject as slippery and as filled with unknowns as physics. But the question intrigued me. How do you make a real person a character? I never met Oppenheimer and yet he, or a version of him, is someone I know intimately, someone who lived inside my head for months. Now it may be, in an existential sense, that we always make people up — how do we truly know someone else? But in this case, the question was more pragmatic: how do you put him on a page?

When did I start imagining him? Another pragmatic answer: it was the summer of 1995, almost exactly this week in June, by the way, and I'd come to Los Alamos as a tourist. I spent hours in Fuller Lodge, looking at drivers' licenses with anonymous numbers and passes and old photographs — what are now the personal artifacts of the Project — and what fascinated me most was the secret nature of the Project. Outside, walking around Ashley Pond, I felt I was in an almost prototypical American town, a place just like anywhere else, and yet in 1945 this was the most secret place on earth. Technically and officially, it did not exist. If you signed on for the Project you literally left the rest of the world behind, communicating through box numbers. What was that like? And it was at that moment that, as the cartoons would have it, a little light bulb went off over my head. What would have happened, I wondered, if there had been a crime? How would they go about solving it? There wouldn't be any police: no one but the Army, and its MPs, was allowed up on the mesa. No one, in fact, was even supposed to know the town was here. How do you solve a crime in a place so secret it doesn't officially exist? I didn't know then the question would be so intriguing that I would spend a year answering it.

But something else happened that summer that drew me deeper into Los Alamos's history. It was the fiftieth anniversary of Hiroshima and, as might be expected, there was a lot of media coverage, much of it revisionist. Every generation, of course, looks at the world through its own perspective, and inevitably, and justifiably, any re-examination of the bomb now brings with it the baggage of fifty years of ambivalence and outright fear. I grew up under the mushroom cloud, too, and I ducked under desks during air raid drills, the way I think all of us of a certain age did, and those fears and ambivalences were all too familiar. And yet I thought that what I was reading seemed

Fuller Lodge
Courtesy of the Los Alamos Historical Society

skewed, not quite right. It didn't fit with what I'd seen in Fuller Lodge, or with what I knew about the project in general. The scientists had become demonized. The reality, I thought, was so much more complicated and interesting. The average age of the scientists on the project was 27: Demons? Or just smart, ambitious kids, who thought they were doing the right thing, for the right reasons? And yet created such appalling consequences. The idea took hold of me and wouldn't let go. What was it like for them? It's very hard for us now to capture the hope and urgency and patriotism of the Manhattan Project — it was, as they say, a different time. But fiction demands a peculiar form of empathy and it often begins, as it did here, with the simple question: what if it had been you? What if you had been the 27-year-old scientist, a kid from Caltech perhaps, and someone had said, "We want you to come to work for the government on a secret project. You will be working with the finest minds in your field. You will cross a frontier in science. You will win the war." I realized that I would have said "Yes," would have been one of the people who helped make the bomb. And who later would have wondered what I had done.

So I was intrigued by the secrecy, then fascinated by the moral ambiguity, and began writing the book in my head. What had it been like? And the Los Alamos Historical Society (to whom all thanks and praise) luckily had exactly the sort of material I was looking for: not diagrams of implosion lenses, but descriptions of the housing units (and the order of their desirability), the utility bills (how much did it cost a month?), the coal deliveries, all the stuff of daily life that really forms the back story to any work of fiction and indeed to any understanding of a community.

Sometimes you can see things in a statistic. When Los Alamos is selected at the end of '42 as the site, one of the reasons, of course, is that the boys' school provided an existing infrastructure — the water tower, some buildings — for what is anticipated to be about 500 people. By the spring of '45, when the book is set, there were 5,000 people on the mesa, and simply by looking at these numbers I could imagine the hill as a constant building site: lumber trucks rolling in the background, carpenters hammering all day long, activity that never stopped. The statistic that was mentioned earlier this morning about the growth of the nursery in itself gives you a clue to who was filling those new housing units — young married couples, for whom the Hill was often their first home. And sometimes a scene will fall right into your lap. I read somewhere that Indian maids were bussed up the mesa twice a week — evidently they were not considered a security risk — and I thought, in the most secret place on earth, there was maid service? What novelist would resist?

So the light bulb became a story and I started my own secret project. I never told anyone I was writing it, in part because I had never written before and I was a publisher — and what could be more embarrassing than a pub- lisher who couldn't write? And certainly at that stage I had no

Laundry at Los Alamos

intention of writing about Robert Oppenheimer. I have always had mixed feelings about the use of real people in fiction. Aside from anything else, read- ers bring their own ideas about them to the page — they already know what they feel. And in this particular case, the last thing I wanted to do was in any way trivialize or misrepresent a figure I respected and admired. I thought enough mud had been thrown at Oppenheimer during his life — I didn't want, even inadvertently, to add to the damage. Nevertheless, in the story someone has to authorize an investigation and it seemed to me as silly to pretend that the Manhattan Project director was, say, "Joe Dawes" as it was to pretend that Franklin Roosevelt wasn't President. It couldn't be anybody but Oppenheimer.

So I took a small breath and thought: right, I only need him for one scene. Do no harm, make it as innocuous as possible, and move on. And then, just as I think happened in real life with Oppenheimer, the moment he opened his mouth, he took over. He started talking on the page with an ease that none of the other characters had for me and one scene led to another — I couldn't get him to stop. I realized that he had become not just a major character in the story but the key to what I really had wanted to write about all along, the story behind the entertainment, the thriller front, of the Project in all its contradictions. For me, as for so many others, Oppenheimer *was* the Project. As was once said of Teddy Roosevelt, "He was the groom at every wedding." It is, of course, unfair to embody an entire enterprise like this in one person, but Oppenheimer's personality (as I think everyone has indicated here today) is a fatal attraction. It's so complicated that you feel that if you can just get him right, you can understand the rest a little.

But how do you get him right? How do you imagine Oppenheimer? The least satisfactory route, ironically enough, was his own words. The letters (recently reissued by Stanford) are certainly wonderful and interesting, but he was famously elusive and ambiguous, as we know, and never more so than in his writing. (And perhaps bored — his only "B" at Harvard was Freshman Comp.) Partly this is a function, I think, of his quick, mercurial mind. He was too fast for prose. He would shift, he would change his mind. He was too smart *not* to change his mind. Like real mercury, he was fluid, not fixed. Later, in the hearings, this would have a devastating effect. "What did you think in 1947? In 1949? In 1951?" Well, he thought different things. Didn't you? I would have. But at a hearing this was considered inconsistent, and perhaps duplicitous.

Sometimes the letters are less than useful because of his own myth-making — he was probably as guilty of this as most famous people are. One of the most quoted of all Oppenheimer lines, of course, is what he said went through his mind after the test at Alamogordo, the line from the *Bhagavad-Gita*: "I am become Death, the destroyer of worlds." It's a wonderful line; the only problem is that the first time he ever said he thought this was three years later, in 1948. He may well have thought it at the time, but what his brother Frank tells us he actually *said* was, "It works."

Of course, because he was so famous, getting all the outside details right was easy. We know what he looked like, how he spoke, how he dressed, even how much he smoked — as someone said earlier, four to five packs a day. Indeed, apart from Einstein, he was probably the only scientist the general public knew at all. His hat became an icon of the period. So there were photographs, seemingly countless photographs, always useful sources. If you're imagining Oppenheimer, you can spend hours looking at these. Like many charismatic people, Oppenheimer photographed very well; the camera loved him. I have a copy of an Eisenstadt photo at home that somebody gave me, and I've kept it, not just because of the unexpected role that Oppenheimer came to play in my life, but because it's a great picture. I wouldn't say that when you look into his eyes you can see his soul, but you can definitely see something. The photographs taken on the hill, at parties, in casual groups, are particularly interesting and poignant, with none of the strain and bleak disappointment that sometimes appears in the later years. He may have been busy and exhausted and under pressure, but in these pictures he always looks as if he's having the time of his life. And, of course, he was.

Even more useful was what other people said about him, and they all said something. I don't think I've ever read anything about Los Alamos in which he doesn't appear, a mass of character detail to draw on and use. People agreed his eyes were mesmerizing, so I used that. Someone said that he had the gift of intimacy — when he spoke to you, he made you feel you were the only person in the room — so I used that. I also used what I imagine must have been its opposite, that when he stopped talking to you, you felt somehow cast out, no longer in the light.

More specious as a technique, I'm afraid, but one that writers use all the time, is drawing what you can from your own life. It so happens that Oppenheimer and I had gone to the same universities, first Harvard and then Cambridge (although it need not be said that I wasn't at the Cavendish Lab), so I knew a little bit about what he'd experienced there. I had run a company for many years, and so I had a fair idea of how much of his day must have been spent in meetings, and dealing with personnel: juggling people, juggling appointments, trying to please everybody and sometimes pleasing no one, not even you. Practically nobody writes about Oppenheimer as an administrator, as a desk man, but he was, and he was good. I think it's probably the reason his prickly relationship with General Groves worked as well as it did. Groves admired anyone, even an intellectual, who could get things done.

But finally, in then end, to imagine Oppenheimer, you have to let the real one go. Fiction can never have the complexity of real life, and who could have made up Oppenheimer? We've heard all the adjectives today: brilliant, arrogant, brittle, empathetic, self-doubting, proud of what he'd accomplished, dismayed at what he'd accomplished. Maybe he was myth-making when he said, "I am become Death," and maybe he did think it, when that first extraordinary flash of light ripped through the sky. The point is that we can believe he did. This is a man who wrote in 1964 to Max Born, his former mentor, and now a public opponent of nuclear weapons, "I have felt a certain disapproval on your part for much of what I have done. This has always seemed to me quite natural, for it is a sentiment I share." Did he mean it? Maybe yes, maybe no. I think, probably both. It was Scott Fitzgerald who said that the test of a first-rate intelligence is the ability to hold two opposing ideas in the mind at the same time, and still retain the ability to function. Who better, then, to hold two ideas at the same time?

He is, in short, a great character — larger than life even in real life. Fiction can't take him in. When you invent Oppenheimer, what you have to do is settle for a piece, or at the most, a few pieces of the puzzle. But fiction does have one great advantage: you get to choose the pieces you want. My Oppenheimer is the Los Alamos Oppenheimer, when his life was as triumphant as the Project itself. A time not without crises or doubts, but not the frightened, vindictive world he would inherit after he made his bargain with the devil.

Oppenheimer and Dorothy McKibbin at
Los Alamos
Courtesy of the Los Alamos Historical Society

That devil was not, I think, atomic energy, for all its terrifying qualities, but the more familiar one: fame and power. The devil for Oppenheimer was in Washington, a place with no room for his dazzling, supple intelligence. Different skills were required there. Having fathered the bomb, Oppenheimer wanted to be its conscience. He had been a hero of one war, but there was no place for him in this new one. He could have been shown the door quietly. Instead he was publicly humiliated, in a way that now looms almost as large in the Oppenheimer legend as the Los Alamos years, and which still resonates in the scientific community.

There is about Oppenheimer's later years that puzzled, the slightly dazed quality of any smart little boy who discovers there are bullies in the playground. Worse, that they are running it. Oppenheimer had often been envied — now he found he had real enemies, and not even his intelligence and achievement could protect him against them. He had expected to remain a world figure. Instead, he was set adrift into a profound disappointment. Maybe it shouldn't have mattered to him, but it did. Maybe the last years needn't have played out as they did — his troubled wife Kitty, lost in her own downward spiral; his reputation growing more and more distant; his Fermi Award (bitter irony) received while still denied access to material he had helped create. Oppenheimer was never to be officially rehabilitated. But time has a way of sorting things out. Does anyone now remember Lewis Strauss as someone other than the playground bully who had it in for Oppenheimer? How many will attend his centennial?

Martyrdom is now an inevitable piece of the Oppenheimer character, but not of the one I made up. The Oppenheimer I imagined does not yet know what is to come, even if we do. He is forever in these pages the king of the hill, the groom at every wedding, the person you want to talk to at the party. He knows this is his chance and he is reaching for greatness. And finding it. He is accomplishing the impossible, he is winning the war, and he is creating a complicated legacy for himself and for all of us. It wasn't difficult to imagine this Oppenheimer. He is as real, I think, as any of them, and he's certainly the one I want to remember.

CHAPTER TWO

Life at Los Alamos

PRESERVATION ON THE PAJARITO PLATEAU

Stuart Ashman
Director, Office of Cultural Affairs, State of New Mexico

I'd like to highlight the historic significance of the Manhattan Project and the activities that took place here in the middle of the last century. But before that, I'd like to relate something that happened to me which exemplifies the kind of place this is. On my way to the Symposium, I stopped at a convenience store. I suppose the clerk knew I was coming to the conference by the way I was dressed, and he volunteered that his father worked at the lab from 1947 to 1953, part of that time alongside Enrico Fermi. Imagine: a convenience store clerk in a small town!

Stuart Ashman

As a former art museum director, I wanted to point out that it is widely accepted in our world that the detonation at the Trinity Site was a turning point for New Mexico as an artistic center. No longer was this place an isolated oasis inhabited by indigenous peoples. It was now part of a bigger world, and had a new reality, artistically, and in every other way. There was a great dichotomy here between the worlds that got thrown together and resulted in the Manhattan Project. Not just politics and science, not just fascism and democracy, but the two separate worlds that coexisted right here in Los Alamos before the development resulting from the Manhattan Project.

The site selected for the top secret research on the atomic bomb combined two lifestyles, two separate worlds. On the one hand, the government in 1942 took over the Los Alamos Boys' Ranch School, one of the most exclusive boarding and prep schools in the United States. In one fell swoop, the government ended Ashley Pond's dream and the school's 26-year history of playing host to the sons of some of the Eastern seaboard's most prominent industrialists.

At the same time, it also forever ended a seasonal migratory lifestyle practised by three generations of Hispanic farmers. Each summer, farmers who wintered in the Rio Grande would migrate to the Pajarito Plateau for the cooler temperatures at the 8,000-plus feet to grow summer crops and raise cattle. The roads, they built for the biannual migrations and for hauling products to market traverse deep canyon walls and rugged arroyos. The routes still can be seen today as narrow, rough roads built on embankments to accommodate them as they crossed the high mesa and descended into the narrow, steep canyons that bisect the landscape.

At the Ranch School, under Pond's philosophy of building men from boys by exposing them to harsh conditions and manual labor, the sons of these industrialists built a second set of roads. These roads were built for horseback riding and provided access to remote areas as part of Pond's "learning by doing" philosophy. The trails were more notable for their craftsmanship, featuring switchbacks, and sophisticated embankments made with rocks cut and fitted by the students. Along with the wagon roads of the migrating homesteaders, they may have provided a framework for transportation routes during the construction of the Manhattan Project.

Although the lifestyles of the migratory farmers and the wealthy businessmen of tomorrow were disparate, they coexisted and provided support for one another. The farmers grew crops and raised cattle, and sold these to the Boys' School, providing them with produce, dairy and meat. The proceeds paid the farmers' taxes and enabled them to continue their tradition of migrating in summer to the plateau. The contribution that the roads and trails made to New Mexico's history and eventually to that of the nation were recognized last fall, when they were listed in the National Register of Historic Places.

Of the 800 acres of ranch property and 2,900 acres of homestead property taken over for the Manhattan Project, few historic resources remain to tell the story. Fuller Lodge, the headquarters, recreation center and guest house for the Ranch School, and many of the trails tell part of the story. They are still used today and are cherished by the community. So are the buildings known as "Bathtub Row." This string of small cottages were used as faculty housing at the Boys' Ranch and got their nickname because of their superior plumbing facilities! They later housed the scientists who built the atomic bomb.

There are other resources on the Hill that tell the story of the Manhattan Project and the ensuing Cold War era that still exist and are in use today. The Los Alamos Post Office was built just after the war as part of a multimillion-dollar community center funded by the Atomic Energy Commission. Much of the community center was subsequently altered, but the post office with its handsome thunderbird grills and large vertical window bays continue to tell the story of Los Alamos after the war. The architect, W.C. Kruger, went on to design the Roundhouse, the state capital in Santa Fe.

Pueblo Indians of New Mexico
Courtesy of the National Archives and Records Administration

The history of the Pajarito Plateau goes back thousands of years before the development of the roads. Evidence of human occupation on the Plateau dates as far back as Paleo-Indian times, to approximately 8,000 B.C. Spear points made and used by ancient hunters following large game animals have been recovered from lands managed today by Los Alamos National Laboratory, as well as other locations on the Pajarito Plateau. The ancestral Puebloan people developed a farming culture and built extensive villages throughout the many canyons and mesas. The density of archaeological sites on the Plateau is the highest in the state of New Mexico.

The descendants of the Pueblo people had a diverse and flourishing cultural tradition by the time the Spanish arrived. And they have survived centuries of pressure from other groups while maintaining their true identity. These are, of course, the Pueblo Indians of northern New Mexico, people today that we know as friends, neighbors and co-workers.

Los Alamos Post Office
Courtesy of the Los Alamos Historical Society

Los Alamos is home to several Lustron houses. After World War II, the U.S. faced a severe housing shortage. Carl Strandlund, a Swedish immigrant who manufactured porcelain enamel steel panels, came up with the innovative idea of using the panels to create prefabricated "maintenance free" houses made of steel. From 1948 to 1950, approximately 2,500 Lustron pre-built homes were trucked to communities across the country. Approximately 47 are individually listed in the National Register, and five states have multiple listings of Lustrons. Because of their scarcity, the Southwest region of the National Trust for Historic Preservation listed them on this year's Eleven Most Endangered List. Last weekend there was a two-day national conference in Columbus, Ohio, that voted to list these, and there are six of these houses on Fairway and 44th Street here in Los Alamos. The Historic Preservation Commission is looking into listing them in the State and National Registers.

The 2000 Cerro Grande fire destroyed many homes, disrupted lives, and took along with it numerous cultural resources. Among them were many of the homesteader cabins that dotted the Pajarito Plateau. Today, only one remains in section C2 of the Los Alamos National Laboratory. Listed last year at the New Mexico Register of Cultural Properties, it was built in 1920 by a descendant of a soldier who served with Don Juan de Oñate around 1600.

The Historic Preservation Division hopes to work with the Lab to preserve a portion of its Cold-War-era past, as it makes way for construction of a new campus. The State Historic Preservation Officer Katherine Slick and her staff toured the Lab earlier this month in light of plans to demolish the Administration Building which was designed by Skidmore, Owings and Merrill, currently the third largest architectural firm in the U.S. The firm was recently commissioned to construct the freedom tower at the World Trade Center site. The Historic Preservation Division is also encouraging the Los Alamos National Laboratory to preserve archival photographs to document the broad range of activities conducted at the Administration Building during this period of significance and to document its original design, so it can be compared to the as-built facility.

The Historic Preservation Division would like to see the Health Research Laboratory retained, as well as landscape elements from the era that speak to its past and current mission. The Division also hoped to be able to comment on the overall master plan for the new campus at the Materials Science Laboratory, so an interpretation of the Health Research Laboratory is included at an appropriate location.

In conclusion, the Department of Cultural Affairs is the primary steward of the State's cultural patrimony. Through our Department's Historic Preservation Division, we have a duty to protect historic buildings and sites for future generations. Los Alamos was a stage for a pivotal period in United States history. The elements of New Mexico history and the impact of the Manhattan Project on it should be preserved and made available for people to observe and interpret for decades to come.

J. ROBERT OPPENHEIMER AND THE STATE OF NEW MEXICO: A RECIPROCAL RELATIONSHIP

Ferenc Szasz

Professor, Department of History, University of New Mexico

Ferenc Szasz

The sagas of famed atomic physicist J. Robert Oppenheimer (JRO) and the state of New Mexico have long been intertwined. Although the scientist lived the majority of his life in New York, California, and New Jersey, the Oppenheimer family leased and subsequently purchased a second home in the northern Pecos Valley during the late 1920s. A generation after his death, Oppenheimer's name still echoes throughout the region. Los Alamos boasts an Oppenheimer Avenue, and the J. Robert Oppenheimer Memorial Committee sponsors a prestigious lecture series that brings speakers to the Hill from around the world. In 1983, on the fortieth anniversary of the founding of the community, the Los Alamos National Laboratory (LANL) renamed their scientific library the "J. Robert Oppenheimer Study Center" (currently the third largest library in the state). Today the scientist's son Peter lives quietly in the Santa Fe region.[10]

---

[10]With the end of the Cold War, the range of speakers at the lecture series now includes eminent Russian scientists such as the former director of Arzamas-16. See the pamphlet *Academician Yuli Borísovich Kharíton*, July 1995 (Los Alamos, NM). Copy generously sent by Roger Meade, Archivist of the Los Alamos National Laboratory; *Los Alamos Monitor*, 17 April 1983, as found in Fern Lyon and Jacob Evans, *Los Alamos: The First Forty Years* (Los Alamos Historical Society, Los Alamos, NM, 1984), p. 171.

Even in the early twenty-first century, the name "Oppenheimer" still calls forth a flood of contradictory images. Throughout his 62 years, JRO wore a number of hats. He was a child prodigy, a Harvard polymath, a pioneer in the emerging field of theoretical physics, the man who put West Coast physics on the world map, and the famed director of the secret Los Alamos Scientific Laboratory (LASL) from 1943–1945. After the war, reporters termed him "the father of the atomic bomb," and for almost a decade, he carried the mantle of public voice of nuclear wisdom. In 1947 he became director of the Institute for Advanced Study at Princeton and, several years later, suffered the indignity of a witch-hunt "trial" before the Personnel Security Board of the United States Atomic Energy Commission (AEC). Because of his well-known left-wing connections in the 1930s, plus his recent opposition to the creation of a U.S. hydrogen bomb, the AEC deprived him of his security clearance. Nine years later, however, the AEC partially apologized by awarding him its prestigious Fermi Medal. In 1994, 27 years after Oppenheimer's death, a former Soviet spymaster, Pavel Sudaplatov, publicly claimed that JRO (as well as several others) had delivered atomic secrets to the Soviet Union during the war. But Sudaplatov produced no documentary evidence to support the claim, and the charges were hotly denied by both historians and JRO's colleagues.[11] A British commentator once compared JRO's post-AEC hearings treatment to that of the infamous Alfred Dreyfus affair, but the template reaches well beyond the anti-semitic France of the 1890s. Indeed, the saga of his life seems ripped from the pages of Sophocles or Aeschylus: J. Robert Oppenheimer as tragic hero of the early nuclear age.[12]

---

[11] Pavel Sudoplatov, *Special Tasks: The Memoirs of an Unwanted Witness — A Soviet Spymaster* (Little Brown, Boston, MA, 1994). *Time*, 25 April 1994, 65–72; *New York Times*, 19 April 1944; *Albuquerque Tribune*, 18 April 1974.

[12] For all his significance, Oppenheimer has been the subject of relatively few biographies. See Peter Goodchild, *J. Robert Oppenheimer: Shatterer of Worlds* (Houghton Mifflin, Boston, MA, 1981); James W. Kunetka, *Oppenheimer: The Years of Risk* (Prentice-Hall, Englewood Cliffs, NJ, 1982); and Jack Rummel, *Robert Oppenheimer: Dark Prince* (Facts on File, NY, 1992). The latest is Jeremy Bernstein, *Oppenheimer: Portrait of an Enigma* (Ivan R. Dee, Chicago, IL, 2004). Kai Bird and Martin Sherwin's long-awaited *American Prometheus: The Triumph and Tragedy of J. Robert Oppenheimer* (Knopf, 2005). Robert F. Bacher's brief pamphlet *Robert Oppenheimer, 1904–1967* (Los Alamos Historical Society, Los Alamos, NM, 1972/1999) is still valuable. The lengthy AEC hearings are available in *In the Matter of J. Robert Oppenheimer: Transcript of the Hearing Before Personnel Security Board and Texts of Principal Documents and Letters* (The MIT Press, Cambridge, MA, 1971). JRO appears as a central figure in Herbert York, *The*

Los Alamos Ranch Trading Post
Courtesy of the Los Alamos Historical Society

International figure though he may have been, J. Robert Oppenheimer also directly shaped the course of New Mexico history. In fact, the JRO/New Mexico stories overlapped in a myriad of ways. Over the years, Oppenheimer forged a number of links to his adopted state. From his first visit to the Pecos Forest Reserve in 1922, to the purchase of a summer retreat in the upper Pecos Valley in 1929, to his directorship of the Lab during the war years, to his honorary degree from the University of New Mexico in 1947, to his poignant last speech in Los Alamos in 1964, the sagas of man and state have long been intertwined. In a strange sense, the relationship proved reciprocal. The natural wonders of northern New Mexico helped restore J. Robert Oppenheimer to both mental and physical health, and his scientific management catapulted Los Alamos into a city that is now recognized around the globe.

Born into a wealthy New York German-Jewish family on April 22, 1904, young Robert's first contact with New Mexico (like so many of his generation) came because of illness. For all his brilliance, JRO suffered from a variety of health problems. His parents cautioned his young friends to simply let him alone during his periodic bouts of depression. His high school classmate and long-term friend Francis Fergusson even claimed that JRO had once tried to strangle him with a belt.

Shortly after he graduated from New York City's famed Ethical Culture School, JRO contracted a severe case of trench dysentery while on a mineralogical trip to the Herz Mountains of Germany. Too ill to attend Harvard that fall as planned, he spent most of the year in the family apartment on Riverside Drive, where he was largely confined to his room: the prescribed treatment of

Advisors: Oppenheimer, Teller, and the Superbomb (W. H. Freeman and Company, San Francisco, CA, 1976), Gregg Herken, *Brotherhood of the Bomb: The Tangled Lives and Loyalties of Robert Oppenheimer, Ernest Lawrence, and Edward Teller* (Henry Holt and Company, NY, 2002), and S.S. Schweber, *In the Shadow of the Bomb: Oppenheimer, Bethe, and the Moral Responsibility of the Scientist* (Princeton University Press, Princeton, NJ, 2000). The Preliminary Proceedings of the Atomic Heritage Foundation's symposium, *Oppenheimer and the Manhattan Project*, (25–26 June 2004), are available in typescript form from the Los Alamos Historical Society.

the day. This would have been hard on any teenager but for a budding genius it approached the impossible. He increasingly began to behave in a sullen and boorish manner. In desperation, his father begged Herbert W. Smith, his former English teacher, to take him West during the summer of 1922 to try to restore his equilibrium. For almost two months the two rode through what is now the Pecos Wilderness of northern New Mexico (then called the Pecos Forest Reserve). A guest ranch in Cowles, New Mexico run by Winthrop and Katherine Chaves Page served as their base.[13]

Ruins of the old church at Jemez
Courtesy of the National Archives and Records Administration

That Smith and Oppenheimer ended up in the mountains of New Mexico rather than those of Montana, Wyoming, or Idaho may largely be credited to fellow Ethical Culture student and Albuquerque resident Francis Fergusson. Fergusson belonged to one of the state's most eminent families. His mother was Clara Huning, daughter of pioneer Fritz Huning, whose Albuquerque mansion — the Huning Castle — once ranked among the state's most elegant homes. His father was lawyer Harvey B. Fergusson, who served in Washington, DC, as both Territorial delegate (from 1896 forward) and, after statehood in 1912, as its first elected Congressman. Francis' older brother, Harvey, later became a respected Southwestern novelist, while his older sister Erna gained even greater fame as the chief interpreter of the region to outsiders. Her Santa-Fe-based Koshare Tour Services (1920–1927) and her books — especially *Our Southwest* (1940) — introduced thousands to the famed "three cultures" of the area. Francis had attended Albuquerque schools, but he transferred to the Ethical Culture School for his high school senior year to better prepare for admission to an Ivy League university.

---

[13] Alice Kimball Smith and Charles Weiner (eds.), *Robert Oppenheimer: Letters and Recollections* (Harvard University Press, Cambridge, MA, 1980), pp. 7–10. Charles Weiner interview with Herbert W. Smith, 1 August 1974. Copy deposited at the American Institute of Physics, College Park, MD.

Francis and Robert shared a schoolboy interest in poetry and drama and surely it was he who urged the Oppenheimer family to consider the healing charms of his home state. Francis obviously knew a good deal about the Koshare Tours, where elegantly dressed young women, decked to the nines with Indian jewelry, escorted La Fonda Hotel visitors to nearby Indian pueblos and ancient Native and Hispanic sites. Through family connections, the Fergussons also had links with the Chaves/Page clans.[14] Moreover, ever since the late 19th century, New Mexico had earned a deserved reputation as a place to regain one's health. During the pre-antibiotic days of the 1930s, the fledgling *New Mexico* magazine frequently ran articles touting the state as a "land of almost perpetual sunshine."[15]

Mountain ranges around Los Alamos
Courtesy of the Los Alamos Historical Society

Oppenheimer's first lengthy visit to New Mexico in the summer of 1922 had a number of consequences. First, Katherine Chaves Page (a 28-year-old upper-class Hispanic woman blessed with great charm of manner) welcomed the frail, insecure boy into her family circle. Women often responded to JRO — on a variety of levels — and some historians argue that he had his first schoolboy crush on her.[16] Although perhaps overstated, Herbert W. Smith later told historian Alice Kimball Smith that when the warm, aristocratic Chaves family embraced the frail Oppenheimer, he found himself loved and admired "for the first time in his life."[17]

Second, JRO's long rides through the Pecos Forest Reserve — still one of the most spectacular regions of the state — helped restore his mental and

---

[14]"Erna Fergusson," *Albuquerque Review*, 8 February 1962; *Albuquerque Journal*, 12 February 1962; *ibid.*, 5 March 1954; *ibid.*, 7 March 1965.

[15]C.H. Gellenthien, M.D. with Anna Nolan Clark, "Climate for Health," *New Mexico* 15 (September 1937) 12. See also Gellenthien with Clark, "Climate: The Magic Difference," *New Mexico* 15 (November 1937) 14–15, 41.

[16]Goodchild, *J. Robert Oppenheimer*, pp. 14–15.

[17]Quoted in Smith and Weiner, *Letters and Recollections*, p. 10.

physical balance. The young men camped amidst mountain forests of spruce, pine, piñon, and aspen. They rode by 13,000-foot peaks, such as Santa Fe Baldy and Pecos Baldy, that stretched well above timberline. They criss-crossed the high meadows of Hamilton Mesa, Round Mountain, and Grady's Mountain, which dazzle the eye every summer with wild hollyhocks, red skyrockets, blue-bells, jack-in-the-pulpits, purple asters, shooting stars, mountain pinks, and bluewood violets. The area also abounds with lupine, blue bonnets, butter-cups, dwarf lobelias, coreopsis, columbines, thistles, and evening primrose. The forks of the Rio La Casa River contain 40'–60' waterfalls, while bear, mountain lion, deer, and elk all drink at sundown from the region's scattered lakes.[18] As nature writer Lou Hernandez once observed, "In spring a visitor can enjoy the feeling that he is the first to set foot in a virgin wilderness..."[19]

Smith later reported that JRO relished the challenges of this mountain experience and accepted his responsibilities like a mature adult. The trip must have helped restore both body and soul for Oppenheimer entered Harvard in the fall of 1923 and completed the rigorous four-year curriculum in only three years. After graduation, he studied in Cambridge, England before earning his Ph.D. in theoretical physics at Goettingen University in 1927, when he was only twenty-three. His performance at the oral examination proved so dazzling that his professor, James Franck, jested, "I got out of there just in time. He was beginning to ask *me* questions."

By switching to theoretical physics, young JRO finally discovered his calling. After two years of further study, he accepted simultaneous appoint-ments at the University of California, Berkeley, and the California Institute of Technology in Pasadena. Alternating semesters at each university, during the next decade Oppenheimer helped put West Coast physics on the international scientific map. Initially apolitical, from the mid-1930s onward JRO increas-ingly moved amidst radical left-wing circles. Whether he officially joined the Communist Party remains an issue of some dispute.[20]

---

[18] For descriptions of the region see Roy Allen Stamm, "Jaunt in July," *New Mexico* (March 1937) 16–17, 34–35; "Trail Riders Plan Trek," *New Mexico Magazine* 27 (March 1949) 26; and Stamm, "The Peaks of the Pecos," *New Mexico Magazine* 5 (September 1937) 22–23, ff.

[19] Lou Hernandez, "High Country Waterfalls," *New Mexico Magazine* 45 (August 1967) 3.

[20] For a vigorous defense of his scientific accomplishments, see John S. Rigden, "J. Robert Oppenheimer: Before the War," *Scientific American* 273 (July 1995) 76–82. Historians Gregg Herken and Barton Bernstein argue that he was a Communist; Martin Sherwin and Kai Bird do not agree. See Charles Burress, "Expert: Oppenheimer was Communist," Associated Press

Once back in the states for good, JRO's father purchased a summer home for his sons in the upper Pecos Valley in 1929. Thus, JRO and his younger brother Frank (also a physicist) shared a New Mexico summer retreat to which they would escape as often as possible. The Oppenheimer brothers grew to love the region, and this also allowed Oppenheimer to maintain contact with Katherine Chaves Page until her tragic murder in 1961. It is no exaggeration to say that the foremost impact that New Mexico had on JRO was to restore him back to health.

Oppenheimer's 1922 trip to New Mexico also expanded his circle of friends to include his first westerners. He and Smith renewed contact with Francis Fergusson in Albuquerque, and there he met Harvey and Erna, as well as their friend Paul Horgan. Horgan, who would later gain international fame as both novelist and historian, formed a close relationship with JRO. He visited the Oppenheimers' Long Island summer home for extensive stays at least twice and he and JRO (who at that time voiced serious literary aspirations) enjoyed themselves immensely. On one visit, the two were out sailing on JRO's sloop the *Trimethy* and got caught in a rip tide that carried them far out to sea. When they failed to return on time, Oppenheimer's father dispatched the family yacht for a search mission. In a later interview, Horgan termed the youthful Fergusson–Oppenheimer–Horgan relationship a "pigmy triumvirate." He also confessed that Oppenheimer was the most brilliant person he had ever met. In retrospect, Horgan regretted that their careers had so diverged. Although Horgan taught on the faculty at Wesleyan and Fergusson became a Professor of Literature at Princeton (with JRO nearby), the three seldom saw much of one another.[21]

Perhaps the most remarkable New Mexico friendship that JRO forged, however, came during the summer of 1937. Exhausted from his West Coast teaching, he had returned to the Pecos region for an extended vacation. While horseback riding through the Valle Grande area, he stopped at Edith Warner's modest tea room on the west side of the Rio Grande, near the lone, wooden Otowi Bridge crossing. The daughter of a Pennsylvania Baptist minister, the

---

article in the *Albuquerque Tribune*, 24 April 2004. Herken makes his case in *Brotherhood of the Bomb: The Tangled Lives and Loyalties of Robert Oppenheimer, Ernest Lawrence, and Edward Teller* (Henry Holt and Co., NY, 2002), especially pp. 43–62.

[21] Smith and Weiner, *Oppenheimer*, p. 8. Interview with Paul Horgan by Alice Kimball Smith, 3 March 1976, Institute Archives and Special Collections, MIT Libraries, Cambridge, MA; interview with Francis Fergusson, 21 April 1976, *ibid*.

quasi-mystical Warner had escaped to New Mexico to write essays and eke out a living in the shadow of San Ildefonso Pueblo. From a rented adobe home on Pueblo land, she served as the Otowi "station agent" for the narrow-gauge Denver and Rio Grande Western Railroad (the Chili Line). Translated, this meant that the headmaster of the

Edith Warner's house

Los Alamos Ranch School, A.J. Connell, had hired her to watch over the extensive Ranch School supplies and luggage until a truck could collect the goods on its thrice-weekly, three-and-a-half-hour journey up the lone dirt road to the Hill. The Chili Line railway storage facility consisted of a boxcar.

In addition, Warner operated a small tea shop from her home where she sold gasoline, ice, candy, and sandwiches to the steady stream of tourists who ventured from Santa Fe to the Frijoles Canyon Indian ruins (now the Bandelier National Monument). Warner's pleasant manner, plus her famed recipe for chocolate cake, charmed Oppenheimer, and he never forgot her.

This chance meeting bore fruit six years later. By early 1943, the Chili Line had gone bankrupt, the Los Alamos Ranch School had been taken over by the federal government, and wartime travel restrictions had cut the Santa Fe tourist trade to a trickle. Thus, Warner faced genuine destitution. In a gesture of great magnanimity, JRO, now director of the secret laboratory, Site Y, encouraged her to establish a "reservations only" restaurant to serve special meals — up to ten people at a time — to a select clientele. Because of security concerns, the clientele consisted exclusively of the Los Alamos scientists and their wives.

Both sides benefited from this experience. Hill residents relished the fifteen-mile drive at sunset down to her home by the Rio Grande "where the river makes a noise." They especially enjoyed the simple meals that she prepared on her wood-burning stove for — as she phrased it — her "hungry scientists." Since she and her San Ildefonso partner Tilano drew most of the produce from their extensive garden, the dinners abounded with five kinds of squash, beans, raspberries, and fresh corn. For all this, she charged only two dollars per person and refused all tips. As this modest sum barely enabled her to cover expenses, several Hill wives helped sell her excess produce to their neighbors.[22]

Edith Warner
Courtesy of the Los Alamos Historical Society

Dining on Native American foods before an open fire in a home without telephone or electricity charmed the LASL scientists. The tranquil setting seemed to provide needed respite from their frenzied pace of life on the Hill. Their regular custom not only allowed Warner to survive the lean war years, it also enabled her to forge friendships with some of the finest minds of the day. She became especially close to Alice Kimball and Cyril Smith, Carson and Kay Mark, Edward and Mici Teller, Niels Bohr, Stan and François Ulam, and Robert and Kitty Oppenheimer. (Tilano always referred to the Laboratory director as "Mr. Op.") Fittingly, it was Kitty Oppenheimer who drove down to inform her of the bombing of Hiroshima. Along the way, Warner alerted the scientists and their wives to San Ildefonso traditions, and cautioned them to avoid certain sacred areas when they hiked throughout the region.[23]

Numerous Los Alamos memoirs recall the charm of dining at Edith Warner's home. Reservations became highly sought after as Hill residents relished the good food and quiet conversations. Eventually the steady demand overburdened the frail Warner and she had to reduce her meal offerings from five days a week to three. But she always found time to feed Robert and Kitty Oppenheimer whenever they asked.

After the war, when a new metal bridge across the Rio Grande threatened to disrupt Edith and Tilano's tranquil lifestyle, Los Alamos scientists joined

---

[22] Bernice Brode, *Tales of Los Alamos: Life on the Mesa, 1943–1945* (Los Alamos Historical Society, Los Alamos, NM, 1997), pp. 120–128.

[23] Interview with Alice Kimball Smith by Helen Homans Gilbert, at Radcliffe College (1987). Copy supplied by the Radcliffe Institute for Advanced Study, Schlesinger Library, Cambridge, MA, p. 63; 1945 Christmas Letter, Edith Warner Manuscripts, Angelico Chavez Historical Library, Santa Fe, NM.

with San Ildefonso builders to erect a new adobe house at a more distant location. The building stands today.

Edith Warner did not long survive the war. In spite of treatment at Los Alamos, she died of cancer in 1951. But her story has evolved into a New Mexico legend, one that drew on Hispanic, Native, and Anglo-American traditions. In 1951, sensing that the end was near, she ordered Tilano two years' supply of blue jeans from Montgomery Ward. He died precisely two years later. Whether their relationship was platonic or physical has never been clear, but everyone agrees that it was deep and enduring. A modern observer has called the saga of Edith and Tilano "one of the great love stories of all time."[24]

Taos novelist Frank Waters fictionalized this story in his clumsy *The Woman at Otowi Crossing* (1966).[25] But Warner found her Boswell in Peggy Pond Church's *The House at Otowi Bridge* (1959), a southwestern classic that has never gone out of print.[26]

Oppenheimer's role in this myth has never been given proper credit. Yet it was his chance encounter with Warner during the summer of 1937 that laid the groundwork for the emergence of a New Mexico legend. On a deeper level, Warner's nourishing meals of traditional Native foods provided yet another way by which the charm of New Mexico allowed the pressured scientists to restore their own delicate balance — not just Oppenheimer this time but scores of others as well. As Cyril Smith once remarked to his wife Alice, "You can't possibly talk about Los Alamos without her."[27]

The official headquarters of the Manhattan Engineering District (or "Manhattan Project," the cover name for the American effort to build an atomic bomb), lay initially on the 67th floor of the Empire State Building, and, later, in the new War Building in Washington, DC. But the major scientific and technological work took place at dozens of locations scattered across the nation. Military laboratories and major universities such as the University of California, Columbia, the University of Chicago, MIT, the University of Minnesota, and the University of Rochester all played crucial roles. When

---

[24] Patrick Burns (ed.), *In the Shadow of Los Alamos: Selected Writings of Edith Warner* (University of New Mexico Press, Albuquerque, NM, 2001), p. 31.

[25] Frank Waters, *The Woman at Otowi Crossing* (Swallow Press, Inc., Chicago, IL, 1966).

[26] Peggy Pond Church, *The House at Otowi Bridge: The Story of Edith Warner and Los Alamos* (University of New Mexico Press, Albuquerque, NM, 1959/1960).

[27] Alice Kimball Smith interview.

Major General Leslie R. Groves assumed overall command of this program on September 12, 1942, he tried to bring order to this diffuse enterprise. Groves insisted that all Project activities be compressed to a single goal: to create a combat-ready atomic weapon in the shortest possible time.[28]

From the onset, Groves insisted on absolute secrecy. He did so for two reasons. First, obviously, was to stem any scientific or technical leaks to the Axis powers (or to the Soviets, who were similarly excluded); second was to increase the absolute shock value whenever the weapon first saw combat use. Thus, Groves insisted on a policy of strict compartmentalization: that is, a person should know only enough to perform his or her specific assignment. Only a handful of people knew the overall purpose of the Manhattan Project. Most workers had no idea what the person on the hall above them was working on. Until

General Leslie R. Groves

Hiroshima, many Los Alamos wives did not know what their husbands were engaged in. Even Groves' wife and two children remained completely in the dark.

But numerous scientists protested. Several said they would not move to Los Alamos if they could not tell their wives everything (Groves relented). Others argued that since the many problems they faced were so interrelated, the policy of compartmentalization would actually delay the final outcome. Thus, they insisted that the Manhattan Project create a spot where all issues could be discussed in a no-holds-barred scientific atmosphere.

Groves bent with the prevailing winds and agreed to set up such a venue. But the requirements were strict: the new location had to be isolated from major urban centers, easily protected by Army security forces, convenient to major transportation networks, and (preferably) already on federally-owned land so as to minimize the difficulties of the appropriation of property.[29]

---

[28] Robert S. Norris, *Racing for the Bomb: General Leslie R. Groves, the Manhattan Project's Indispensable Man* (Steerforth Press, South Royalton, VT, 2002) is a first-rate biography.
[29] *LASL NEWS*, 1 January 1963, 13.

Over the objection of many of his advisors, Groves selected JRO to head this new installation. Opponents pointed to Oppenheimer's well-known radical past but Groves argued that Oppenheimer knew so much about the Project anyway that it would be far better to have him at Los Alamos under constant surveillance. Moreover, he sensed that Oppenheimer's relentless ambition for fame would cause him to drop all previous radical contacts.[30]

Although Groves would later claim that he knew well the region of northern New Mexico from his time spent in Arizona, it was clearly Oppenheimer who alerted him to the possibility of New Mexico as a spot for the proposed Site Y. (A California venue had already been rejected as not sufficiently isolated.) In the fall of 1942, JRO and Groves seriously explored two New Mexico locations. The first was Jemez Springs, but Groves felt that the cliffs bordering the town might hamper future expansion, while JRO argued that they would stifle creativity by making the scientists claustrophobic.

The second choice lay with the nearby Los Alamos Ranch School, which by 1942 had fallen on hard times. Many of the staff members had departed for the military, and student enrollment suffered accordingly.[31] One often reads that JRO chose the location because he had graduated from the Ranch School — even Paul Horgan held this view — but this is not so. JRO knew the region only through horseback visits from his Pecos Valley home.

Groves responded favorably. He liked the fact that Los Alamos could be approached only by a single, easily guarded dirt road, and he deemed the Hill sufficiently isolated from mainstream American life. Initially the modest Ranch School buildings seemed appropriate to house the estimated 100 scientists and their families needed to complete the task.

Typical muddy road in Los Alamos, NM

---

[30] Norris, *Racing for the Bomb*, p. 242.

[31] For a superb memoir of life at the school, see John D. Wirth and Linda Harvey Alridrich, *Los Alamos: The Ranch School Years, 1917–1943* (University of New Mexico Press, Albuquerque, NM, 2003). See also Roland A. Pettitt, *Los Alamos Before the Dawn* (Pajarito Publications, Los Alamos, NM, 1972), pp. 42–43.

Although the Ranch School, the nearby Anchor Ranch, and the properties of scores of Hispanic ranchers would have to be appropriated, much of the land lay in Forest Service hands. Oppenheimer also argued that the breathtaking views from the mesas would spur scientific creativity. Both were proven correct. With numerous MP's, many on horseback, a team of guard dogs (the K-9 Squad), and scattered G-2 (Army Intelligence) agents in Albuquerque and Santa Fe, Los Alamos proved relatively easy to secure from outsiders. And numerous scientists recalled the spectacular environment as the lynchpin of their experience. As metallurgist Cyril Smith observed, "to my mind the landscape is as much a part of the project as Groves' management. The environment has an immense effect, I think, on one's general state of mind. Weekend hikes made it possible for us to maintain this intense level of work during the war."[32]

Thus on December 7, 1942, Ranch School headmaster A.J. Connell, who had long seen the handwriting on the wall, received official notice that the Federal government planned to confiscate the Ranch School properties for the war effort. After considerable bickering as to the date of transfer, the Army agreed that the four seniors could take accelerated classes and graduate on January 21, 1943. Accordingly, the four received their diplomas in a formal graduation ceremony and departed, respectively, for Cornell, Harvard, Stanford, and the Newark College of Engineering.[33] The dust had not settled before the Army arrived en masse to create the top-secret Site Y. Without JRO's deep affection for northern New Mexico, the nation's premier weapons laboratory would not be located where it is today.

Oppenheimer's major impact on the state of New Mexico, of course, lay with his role as director of the Los Alamos Scientific Laboratory from 1943–1945. One of the first decisions he made in early 1943 involved trees. Contractors' bulldozers had begun to level the terrain for various buildings but he stopped this immediately, insisting that every possible tree remain in place. This proved among the first of thousands of decisions in what became nothing less than an administrative miracle. JRO's fair and balanced decision-making

---

[32] Interview by author with Alice and Cyril Smith (telephone); copy of transcript in Los Alamos National Laboratory Archives.

[33] *Los Alamos Monitor 50th Anniversary Guide*, Sunday, 28 March 1999, 6. The majority of the text for this special edition came from Marjory Bell Chambers, "Technically Sweet Los Alamos: The Development of a Federally Sponsored Scientific Community", unpublished Ph.D. dissertation, University of New Mexico (1974).

proved even more remarkable when one realizes that he had had no previous administrative experience. (He had never even served as head of a university physics department.)

The issues he dealt with ranged from the mundane to the cosmic. Because of the isolation and relatively primitive living conditions, he spent countless hours assuring scientists — and especially their wives — that the ultimate goal was a worthy one and that the harsh circumstances were only temporary.[34] When Edward Teller refused to cooperate, JRO wisely

"Remote Handling" at Los Alamos
Courtesy of the Los Alamos Historical Society

gave him his own Group to head. Along the way he composed a little prayer: "May the Lord preserve us from the enemy without and the Hungarians within."[35]

The ever-present Army security measures proved a constant annoyance. All incoming and outgoing mail passed under censors' eyes, state driver's licenses carried numbers rather than names, and the numerous babies were registered as born at Box 1663, Santa Fe, NM. Cameras were discouraged and diaries were forbidden. Everyone visiting Santa Fe shops and museums had instructions not to speak to people more than necessary. When Hill wives dined at the La Fonda Hotel for lunch, they were well aware that they remained under constant G-2 surveillance.[36] One final example: the Girl Scouts formed a chapter but the troop had to pretend to be in Santa Fe (as Los Alamos did not officially exist). Moreover, all girls who had scientist fathers were instructed to register under fake names, lest spies discover where their fathers were located.[37]

---

[34] John Marble, "First Medical Staff Couldn't Keep Up . . . ," *Los Alamos Monitor 50th Anniversary Guide*, Sunday, 28 March 1999, 22.

[35] Charles L. Crutchfield, "The Robert Oppenheimer I Knew," in *Behind Tall Fences: Stories and Experiences About Los Alamos at its Beginning* (Los Alamos Historical Society, Los Alamos, NM, 1996), p. 173. The best book on the Manhattan Project remains Richard Rhodes, *The Making of the Atomic Bomb* (Simon & Schuster, NY, 1986). See also the official AEC history, Richard G. Hewlett and Oscar E. Anderson, Jr., *The New World, 1939/1946* (The Pennsylvania State University Press, University Park, PA, 1962).

[36] Gilbert interview with Alice Kimball Smith.

[37] *Albuquerque Journal*, 3 April 1965.

The technical and scientific dilemmas that Oppenheimer faced, of course, dwarfed all the other problems combined. From 1943–1945 Manhattan Project scientists stood at the very edge of human knowledge. This proved true for every step of the process. On a technical level, skilled machinists had to devise delicate beryllium-copper tools to drill high explosives, lest a steel tool accidentally spark and send the building up in flames. Metallurgists dealing with plutonium had no precedent to guide them as they painstakingly shaped the man-made radioactive element into hemispheres for the Trinity/Nagasaki weapons. Among JRO's most creative administrative decisions came with his plan to restructure the entire laboratory to focus on the implosion weapon when it became clear that plutonium could not be used in the planned uranium-235 gun-type bomb.[38]

Along with administrative and technical questions, JRO faced a number of cosmic questions as well. Should the U-235 weapon be dropped in combat without any prior field test? Could a "Super" or hydrogen weapon — hundreds of times more powerful — be created along with a fission bomb? How could the laboratory best defend its workers against potential radiation dangers? And, the ultimate question: could a nuclear bomb somehow ignite the atmosphere and destroy all life on earth? When Los Alamos historian David Hawkins first raised this last question with Edward Teller, Teller responded, with his usual dry wit: "Oh, David, there are worse things that could happen."[39] Oppenheimer had a role in virtually every LASL decision, large or small.

But Los Alamos allowed time for play as well as work. The lab closed down on Sundays and the scientists took advantage of the time off to hike the extensive mountains of the region. They visited the ruins of Frijoles Cañon so often that it became virtually their private playground. In winter they skated Ashley Pond and skied the nearby mountains slopes. Oppenheimer and his wife Kitty took several two-day horseback rides across the Rio Grande to their

---

[38] See Lillian Hoddeson, Paul W. Henriksen, Roger A. Meade, and Catherine Westfall, *Critical Assembly: A Technical History of Los Alamos During the Oppenheimer Years, 1943–1945* (Cambridge University Press, Cambridge, MA, 1993).

[39] Terry L. Rosen, *The Atomic City: A Firsthand Account by a Son of Los Alamos* (Sunbelt Eakin, Austin, TX, 2002), quoted p. 8; on Teller, see Edward Teller with Judith Shoolery, *Memoirs: A Twentieth-Century Journey in Science and Politics* (Perseus, Cambridge, MA, 2001), and Peter Goodchild, *Edward Teller: The Real Dr. Strangelove* (Weidenfeld and Nicolson, London, 2004).

Pecos home, much to the dismay of the FBI agents who had to accompany them. He usually rode a horse named "Crisis," so feisty that he alone could handle it.

All the Los Alamos memoirs laud Oppenheimer's administrative skills. Rather than govern from an office desk, JRO spent his time constantly attending meetings and visiting laboratories. He seemed to know a bit about every problem on hand. People especially remembered his consideration and even-handedness. For example, in one instance he went out of his way to thank the MP's for keeping the project safe. In another case, when he had to choose between two equally qualified scientists for a key position, he simply asked them to draw straws. As Project veteran Jo Ann Foley once observed, JRO should get an award for "cross-cultural communication. He kept everybody on an even keel, even the teenagers."[40]

His administrative reputation has not faded with time. As Atomic Energy Commission head Glenn T. Seaborg observed in 1965, to a large extent "the greatness [of Los Alamos] lay in Robert Oppenheimer."[41] British Mission member James Tuck remained convinced that "a lesser man could not have done it." As another Los Alamos alumnus observed, "The work certainly would have been completed without Oppenheimer, but it wouldn't have been done so soon. He was very close to being indispensable. You think someone else might have come along — but you never know."[42]

Oppenheimer made thousands of decisions that affected New Mexico, but few proved more important than his acceptance of the plan to test the world's first atomic weapon at Trinity Site, about 35 miles east of Socorro. He even chose the name "Trinity" for the spot, and today "Trinity Site" appears on most state maps. What Oppenheimer meant by this term has never been

---

[40] Cited in Ellen D. McGehee, "The Women of Project Y: Working at the Birthplace of the Bomb, Los Alamos, New Mexico, 1942–1946," MA thesis in history, University of New Mexico (2004), p. 106, and "J. Robert Oppenheimer: As Los Alamos Knew Him," *The Atom* (March 1967) 2.

[41] Glenn T. Seaborg, "Los Alamos: 25 Years in the Service of Science and the Nation," *The Atom* 5 (March 1965) 5.

[42] Tuck is quoted in Noel Pharr Davis, *Lawrence and Oppenheimer* (Simon & Schuster, NY, 1968), p. 187; unnamed scientist quoted in Lincoln Barnett, "J. Robert Oppenheimer," *Life* 27 (10 October 1949) 133.

clear. Some have suggested that he drew the name from a John Donne poem he had just read that contained the lines:

"Batter my heart
Three person'd God. . . "

Los Alamos historian Marjorie Bell Chambers, however, has offered another explanation. She argues that the reference to Trinity has Hindu rather than Christian roots. In this sense, the term refers to that which is, is destroyed, and is revived again.[43] The issue will probably never be resolved. When I posed this question to Frank Oppenheimer in the early 1980s, he confessed that he simply did not know.

In the tense hours before the Trinity test on Monday, July 16, 1945, Oppenheimer seemed on the verge of collapse. Gaunt and exhausted, he held onto a pole to steady himself as the Trinity countdown approached zero. During the final seconds, he hardly breathed.[44] When the huge ball of fire rose 40,000 feet in the air (proving that the scientists' theories had been correct) he confessed to a colleague: "My confidence in the human mind is somewhat restored." To his brother Frank he simply said, "It worked."[45] Three weeks later, on August 6 and August 9, the specially modified B-29s Enola Gay and Bock's Car dropped their respective weapons on Hiroshima and Nagasaki. On August 14, Japan surrendered. Although the atomic bombs may not have won the war, they certainly ended it. Under Oppenheimer's direction, the Los Alamos scientists had been given an impossible assignment. And they delivered.

Exhausted by his ordeal, JRO told Groves that he hoped to resign as soon as possible. On October 16, 1945, his last day as director, JRO spoke to virtually the entire Los Alamos community in a gigantic outdoor ceremony. General Groves presented a certificate of appreciation from the Secretary of War, which JRO accepted on behalf of the Laboratory with a brief speech. In it he noted:

> If atomic bombs are to be added as new weapons to the arsenals of a warring world, or to the arsenals of nations preparing for war, then the time will

---

[43] Marjorie Bell Chambers, "Technically sweet Los Alamos: The development of a federally sponsored scientific community," Ph.D. dissertation, University of New Mexico (1974).

[44] General T.E. Farrell's account in Donald Porter Geddes (ed.), *The Atomic Age Opens* (Pocket Books, NY, 1945), p. 32.

[45] For the saga of this event, see Ferenc Morton Szasz, *The Day the Sun Rose Twice: The Story of the Trinity Site Nuclear Explosion, July 16, 1945* (University of New Mexico Press, Albuquerque, NM, 1984/1995).

come when mankind will curse the names of Los Alamos and of Hiroshima. The peoples of the world must unite, or they will perish.[46]

Three weeks later, on November 5, 1945, he gave his final Los Alamos speech to the approximately 500 members of the newly formed Association of Los Alamos Scientists (ALAS). Since he had officially turned the Laboratory over to his successor, Naval Commander Norris Bradbury, he felt a little more free to express his thoughts. Emphasizing both the "peril" and "hope" of atomic energy, he pleaded for scientific openness as the key to world unity. It remains one his best-remembered addresses.[47]

Over the years, many of JRO's scientific colleagues have recorded their impressions of his wartime leadership of the Laboratory. Charles Crutchfield noted that at the start, Oppenheimer seemed to view the Manhattan Project with virtual indifference. He considered it as a purely a scientific inquiry to see if the Allies could crack the nuclear secrets of nature. (If they could not, of course, then neither could the Germans.) But as time wore on, Oppenheimer became more emotionally involved until by 1945 he had staked everything he had on the successful outcome of the Trinity test.[48] Scientist Louis Rosen recalled his former director as a man with the brain of an Einstein and the soul of poet.[49]

Trinity Site Headquarters

Long-term assistant John Manley provided yet another perspective on his leadership during those years when he highlighted JRO's great flair for the dramatic. When JRO began recruiting scientists to come to Los Alamos, it almost seemed as if he were casting them as actors in a play. On virtually every major occasion, he came up with a poignant, quotable phrase. For example, when President Franklin Delano Roosevelt died on April 12, 1945, Oppenheimer quoted from the Hindu sacred writings, the *Bhagavad-Gita*: "Man is a creature

---

[46] Smith and Weiner, *Letters and Recollections*, pp. 310–311.
[47] *ibid.*, pp. 315–325.
[48] Crutchfield, "Oppenheimer," in *Behind Tall Fences*, pp. 172–176.
[49] Rosen, *The Atomic City*, p. 45.

whose substance is faith. What his faith is, he is."[50] After the Trinity test, he delivered an even-more famous quotation from the same source: "I am become death, the shatterer of worlds." (He admired this skill in others, as well. He later told the daughter of Trinity Site Director Kenneth T. Bainbridge that her father's comment — "Now we're all sons of bitches" — was the best thing ever said at Trinity.[51])

This propensity for delivering the on-target, dramatic phrase became a central part of his post-war persona. On August 17, 1945 he said, "A scientist cannot hold back progress because of fears of what the world will do with his discoveries."[52] When he met President Harry Truman, he said he had "blood on his hands." Later he achieved notoriety for the remark: "In some sort of crude sense, which no vulgarity, no humor, no overstatement can quite extinguish, the physicists have known sin, and this is a knowledge which they cannot lose." (Truman was not pleased at such public displays of guilt, and supposedly refused to ever see him again.) In March 1946, JRO advised University of Pennsylvania students that because nuclear war had become "unendurable," the atomic bombs would produce a "better world."[53] In another speech he noted that the "book of the past is closed and one has a fresh page to write on."[54] When asked to characterize his position as director of the Institute for Advanced Study, he described himself as simply an "academic innkeeper." On another occasion, he described the issue of lingering radioactivity in the soil as "a nontrivial problem." After the war he honed this ability into a fine art. In 1949 he noted, "As long as men are free to ask what they must, free to say what they think, free to think what they will, freedom can never be lost, and science will never regress." In 1953 he described the world situation as: "We may be likened to two scorpions in a battle, each capable of killing the other but only at the risk of his own life." Three years later he observed, "In a free world, if

---

[50] FDR Memorial Address, Box 262, J. Robert Oppenheimer Papers, Manuscript Division, Library of Congress, Washington, DC.

[51] Quoted in Robert V. Pound's obituary of Kenneth Thompkins Bainbridge, *Physics Today* (January 1997) 81.

[52] Santa Fe *New Mexican*, 17 August 1945, as found in War Records Library Collection, Scrapbook 71, New Mexico State Records Center and Archive, Santa Fe, NM.

[53] Truman statement, misquoted in Goodchild, *J. Robert Oppenheimer*, p. 174; Santa Fe *New Mexican*, 27 March 1946, Scrapbook 71, New Mexico State Records and Archive Center, Sante Fe, NM.

[54] JRO, untitled tape recording, c. 1947, Audio Division, Library of Congress, Washington, DC.

it is to remain free, we must maintain, with our lives if need be yet surely by our lives, the opportunity for a man to learn anything." On Einstein's death in 1955 he said, "Any man whose errors take ten years to correct is quite a man." When he accepted the AEC award from Lyndon Johnson in 1963, he wryly noted, "I think it is just possible, Mr. President, that it may have taken some charity and some courage for you to make this award today." A number of his sayings have virtually entered the language.[55]

Although JRO may have lacked a sense of humor, he compensated for it by his flair for the piercing *bon mot*. As reporter Eric Sevareid observed in 1963, he was a "scientist who wrote like a poet and speaks like a prophet."[56] And this largely began with his years as LASL director.

In his autobiography, General Groves suggested that at war's end he was eager to see JRO leave LASL for two reasons. First, everything afterwards would be anti-climatic for him and second, Groves expressed concern over the ever-present problem of JRO's radical past.[57] But the exhausted Oppenheimer was more than ready to return to academic life at Caltech, which eagerly welcomed him back. To his dismay, he found this impossible. During 1947, for example, he flew from California to Washington fifteen times. He spent endless hours writing the Acheson–Lilienthal report — the basis for the (failed) Baruch Plan presented to the United Nations in 1946. This proposal embodied America's attempt to avoid an arms race by creating an International Atomic

Life magazine, October 1949

[55] J.K. McCaffery video interview with JRO, Audio Division, Library of Congress, Washington, DC. Many Oppenheimer quotations may be found on http://www.quotations.com and in Clifton Fadiman (ed.), *The Little, Brown Book of Anecdotes* (Little, Brown and Company, Boston, MA, 1985), p. 435. Most compilations of twentieth-century quotations include one or more of his observations, and, incidentally, they rarely cite the same ones.

[56] Eric Sevareid interview with JRO, 2 December 1963, Oppenheimer Papers, Manuscript Division, Library of Congress, Washington, DC. See also, "The Oppenheimer Years, 1943–1945," *Los Alamos Science* 4 (Winter/Spring 1983) 6–25.

[57] Norris, *Racing for the Bomb*, p. 446.

Energy Committee under UN auspices. Unfortunately, the Soviet Union refused to cooperate. Congress also called on JRO on numerous occasions. From director of an obscure secret laboratory in northern New Mexico, Oppenheimer had suddenly assumed the mantle of public spokesman on issues of nuclear science and government.

With the possible exception of Albert Einstein, JRO emerged in the post-war period as the world's most highly profiled scientist. His distinctive porkpie hat — adopted at Los Alamos because Groves felt he stood out too prominently wearing his traditional cowboy hat — appeared without caption on the cover of the inaugural issue of *Physics Today*. *Life* also put him on the cover of its October 1949 issue with the sidebar: "No. 1 Thinker on Atomic Energy."

Given the increased public demands, Oppenheimer made relatively few official trips back to New Mexico after 1945. He did visit the Hill in August of 1946 for a six-day conference on nuclear physics and again the next year to report (favorably) on the status of the now-permanent, Bradbury-run laboratory. We know that he stayed several days with his former secretary Dorothy McKibbin during the time, but his later visits to the state all remain undocumented. In 1947 he assumed the position of Director of the Institute for Advanced Study at Princeton, and in the mid-1950s he and Kitty purchased another summer house in the Virgin Islands. This island home, which allowed him to indulge his passion for sailing, seems to have replaced the Pecos valley cabin as his primary retreat. Although he must have visited his brother Frank's Pagosa Springs, Colorado ranch on occasion, JRO did not make another official public visit to New Mexico until May of 1964.

The fledgling University of New Mexico in Albuquerque, however, had attempted to entice him back in 1947, but without success. From the onset, the academic "manager" of LASL had been the University of California, Berkeley, which oversaw all equipment purchases for the Manhattan Project years. At the dawn of the twenty-first century, with the university LANL contract now open to bids, some have wondered why Groves did not consider the much closer University of New Mexico in Albuquerque for this assignment. One reason for the Berkeley choice surely lay with JRO's long-term links to the school, but another lay with the respective sizes of the two institutions in the 1940s. The University of New Mexico then housed fewer than 100 faculty and under 1,500 students. The 350 graduates who donned caps and gowns in 1947 represented the largest graduating class in the school's 58-year history. (That same year, UNM awarded its first two Ph.D.'s.)[58] From a security point of view, any attempt to funnel gigantic amounts of scientific equipment through UNM in the 1940s would have instantly raised eyebrows. Only an institution the size of Berkeley could have served as an appropriate cover. Indeed, the first Berkeley contract, signed April 20, 1943 and backdated to the first of January, spoke of 250 workers and 7.5 million dollars in expenses.[59] The University of California has successfully managed Los Alamos for over sixty years.

But post-war UNM had its eyes fixed firmly on the future, and in the spring of 1947, President J.P. Wernette wrote JRO to ask him if he could come to Albuquerque to speak at commencement and also receive an honorary degree. Unfortunately, the invitation was delayed in the mails and by the time it arrived, Oppenheimer had made other plans. Still, as he wrote Wernette, "[New Mexico] is almost a home state to me for many reasons... "[60]

Although disappointed the UNM faculty voted to award him the degree in absentia. Accordingly, Wernette read the following at commencement exercises in Zimmerman Stadium on Saturday, June 7, 1947:

> J. Robert Oppenheimer, inspiring teacher, brilliant theorist in contemporary physics, former director of Los Alamos Scientific Laboratory, leader in the development of atomic energy, scientific statesman, determined to make this fabulous power serve the peacetime needs of humanity. Upon the recommendation of the Faculty of the University and by vote of the Regents, I confer upon him, in absentia, the honorary degree of Doctor of Science.[61]

The last thirteen years of Oppenheimer's life, 1954–1967, were not especially pleasant. The publicity surrounding the 1954 AEC hearings, which sullied his reputation, aged him terribly, and seemingly broke his heart. It also made him *persona non grata* in many official circles. A 1955 proffered invitation to speak at the University of Washington was hastily withdrawn, although he did speak to the nearby University of Oregon in Eugene shortly

---

[58]*Albuquerque, Journal* 8 June 1947.
[59]Charmain Schaller, "General Groves Demanded a Miracle—And Got It," *Los Alamos Monitor 50th Anniversary Guide*, Sunday, 28 March 1999, 7.
[60]Wernette to JRO, 11 June 1947, Box 227, Oppenheimer Papers, Manuscript Collection, Library of Congress, Washington, DC; JRO to Wernette, *ibid.*
[61]*Albuquerque Journal*, 8 June 1947.

afterwards. Nuclear politics were so sensitive that Lab director Norris Bradbury could not extend him an official invitation to return to the Hill until after JRO had received the AEC Fermi Award in 1963. With this, Oppenheimer became somewhat "rehabilitated," and Bradbury officially invited him back for a public talk. Oppenheimer chose the subject "Niels Bohr and Atomic Weapons."[62]

On their two-day visit to Los Alamos in mid-May of 1964, Kitty and Robert Oppenheimer received the red carpet treatment. They were treated to a special screening of the documentary film "Ten Seconds that Shook the World," and took a private tour of the newly erected LASL Museum. When JRO sat again in his old director's chair — now a museum exhibit — he quipped to Bradbury that it was "still very hard."[63]

The turnout for his talk proved overwhelming. Although absent from the Hill for almost fifteen years, a thousand people filled the Civic Auditorium that Monday night. The standing ovations that he received at both start and finish brought tears to his eyes. Norris Bradbury introduced him as "Mr. Los Alamos."

During his speech, he emphasized the need for an "open world" of free interchange of scientific information. In passing, he observed that the leaders of the Manhattan Project were "not free of misgivings... We were troubled about what we were up to." Several reporters rushed up afterwards but he refused to respond to any questions. All he would say was: "I love my country, if that's what you want to know."[64]

When Oppenheimer died on February 18, 1967, after months of battling throat cancer, accolades poured in from around the world. Among the most insightful was that of Norris Bradbury. Said his successor: "His stamp upon the character of Los Alamos was profound and permanent; his impression upon those who knew him was no less so... Such men are incredibly rare."[65]

---

[62] Jane A. Sanders, "The University of Washington and the Controversy over J. Robert Oppenheimer," *Pacific Northwest Quarterly* 70 (January 1979) 8–19; "Oppenheimer to Speak Here," *The Atom* (April 1964) 1.

[63] "Los Alamos Revisited," *The Atom* (June 1964) 11–13; quotation on p. 12.

[64] *Albuquerque Journal,* 19 May 1964. A version of the speech appeared in the *New York Review of Books,* 17 December 1964, 6–8.

[65] "J. Robert Oppenheimer," *The Atom* (March 1967) 4.

The links between J. Robert Oppenheimer and the state of New Mexico stretched over four and a half decades, and in many ways they proved reciprocal. The magnificent silences of the Pecos Forest Reserve helped restore Oppenheimer's physical and psychic balance and he developed a lifelong affection for the state. In a sense, he must have linked New Mexico with the idea of renewed health. His consideration for the woman at the Otowi Bridge, whom he met on a 1937 horseback ride, not only allowed Edith Warner to survive the war economically, it also enabled his fellow LASL scientists to retain their delicate mental balance as well. It is no exaggeration to say that Oppenheimer provided the canvas on which Peggy Pond Church would later create an authentic New Mexico legend. JRO's acquaintance with the Los Alamos Ranch School helped convince General Groves to locate Site Y on the Pajarito Plateau. And his directorship of Los Alamos during the war years still serves as the template for effective scientific management.

In 1948, the AEC officially decided to keep the Los Alamos National Laboratory where it was and inaugurated a 100-million-dollar rebuilding program. In 1949, Los Alamos became a separate county. Five years later, the town ranked as the eighth largest city in the state with over 12,000 in population. By the mid 1950s, it represented a 250-million-dollar federal investment. In 2004 its annual budget was two billion.

Passage from A to B building
Courtesy of the Los Alamos Historical Society

The influx of federal monies to a poor region of a generally impoverished state has been without precedent. A 1996 economic survey of the impact of Los Alamos National Laboratory on northern New Mexico concluded that one of every 23 state jobs was either created or supported by the laboratory. The 1.1-billion-dollar funding for the fiscal year 1996 multiplied into about four billion, or about five percent of the total economic activity for the entire state.[66] In an unforeseen way, the spread of this income throughout northern

---

[66]Robert R. Lansford *et al.*, *The Economic Impact of Los Alamos National Laboratory on North-Central New Mexico and the State of New Mexico, Fiscal Year 1996* (Office of Technology and Site Programs, Albuquerque, NM, 1997), p. 13.

New Mexico has allowed traditional Native American and Hispanic crafts to revive and flourish. The Lab provided steady employment for the craftspeople as well as potential purchasers for their various art works.

Yet there is a dark side to the Oppenheimer link with New Mexico as well. Since the Manhattan Project was viewed as a crash program, few gave much concern to the long-term environmental consequences. Its successor agency, the Atomic Energy Commission, did not institute its first committee along these lines until 1947.[67] At the dawn of the twenty-first century, however, environmental issues have come to virtually dominate national and state concerns. The terrible Cerro Grande fire of 2000 allegedly uncovered 300 toxic sites on the Hill. In 2004, the New Mexico State Environment Department, aided by various private organizations, remained locked in conflict with the Laboratory over environmental pollution, especially over potential contamination of the Rio Grande by creeping plutonium.[68]

Not all these environmental problems can be laid at Oppenheimer's doorstep, of course, but he does bear responsibility for a few. In 1975, the successor to the AEC, the U.S. Energy Research and Development Administration (ERDA) discovered a plutonium "pocket" south of the Los Alamos Inn. Further research concluded that the now-open location originally housed the Technical Area laundry. As plutonium washed off workers' clothes, it ended up lodged in the drain. Eventually, the contamination was removed to a burial site.[69]

The main area of radioactive contamination from the Oppenheimer years remains that of the Trinity Site, for the July 16, 1945 detonation surely fell under his watch. Since the Trinity "gadget" exploded only 100 feet above the ground, the ball of fire touched the earth, fusing the sand into radioactive greenish-grey glass and driving the plutonium deep into the soil. Since the half-life of plutonium is about 24,000 years, the fenced-in area of Trinity Site in central New Mexico will reflect the legacy of J. Robert Oppenheimer essentially forever.

A theologian once observed that the concept of forgiveness should lie at the heart of any just society, for one can never anticipate the consequences of

[67] Stephen I. Schwartz (ed.) *Atomic Audit: The Costs and Consequences of U.S. Nuclear Weapons Since 1940* (Brookings Institution Press, Washington, DC, 1998), p. 356.

[68] Laura Paskos, "New Mexico Goes Head to Head with Nuclear Juggernaut," *High Country News*, 24 November 2003, 7–12.

[69] *Albuquerque Journal*, 20 August 1975; *ibid.*, 23 August 1975.

one's actions. Although that observation holds true for the conventional areas of life, because of the extended time periods involved, it seems to resonate with special intensity whenever one speaks of things nuclear. It certainly should apply to Oppenheimer as well.

Although the first generation of atomic scientists overflowed with brilliance, with the passing of years JRO has assumed the highest profile of them all. Similarly, although the Manhattan Project could never have succeeded without the contributions of Oak Ridge, Tennessee, and Hanford, Washington, the community most remembered today is Los Alamos. Perhaps the reciprocal relationship between the state and the man may be summarized as this: the awesome splendor of northern New Mexico restored J. Robert Oppenheimer to health, and, in return, the state now houses both Trinity Site and a magnificent scientific laboratory that is recognized, for better or worse, around the globe.[70] Whether this ranks as a fair exchange depends largely on the perspective of the observer.

---

[70] See the concise summary: "The Legacy of Los Alamos," in Hoddeson *et al.*, *Critical Assembly*, pp. 402–417.

ROBERT OPPENHEIMER: A WINDOW ON HIS LIFE AT LOS ALAMOS

Kai Bird
Fellow at the Woodrow Wilson International Center for Scholars

Martin Sherwin
Professor of History at Tufts University

Kai Bird (*left*) and Martin Sherwin (*right*)
Photo by Claudio Vazquez

(An excerpt from their biography: *American Prometheus: The Triumph and Tragedy of J. Robert Oppenheimer* (Knopf, 2005).)

Los Alamos was an army camp — but it also had many characteristics of a mountain resort. Robert Wilson had just finished reading Thomas Mann's *The Magic Mountain*, and sometimes he now felt he had been transported to that magical dominion.[71] Western civilization was fighting a global war for its very survival, but many of the physicists at Los Alamos were overcome with feelings of sheer exhilaration. It was a "golden time" said the English physicist James Tuck: "Here at Los Alamos, I found a spirit of Athens, of Plato, of an ideal republic."[72] It was an "island in the sky," or as some new arrivals dubbed it, "Shangri-La."[73]

---

[71] Robert R. Wilson, "A Recruit for Los Alamos," *The Bulletin of the Atomic Scientists*, March 1975, 41.

[72] Thorpe & Shapin, "Who was J. Robert Oppenheimer," *Social Studies of Science*, August 2000, 547.

[73] Charles Thorpe dissertation, p. 182; Jane S. Wilson and Charlotte Serber, *Standing By and Making Do: Women of Wartime Los Alamos* (Los Alamos Historical Society, Los Alamos, NM, 1988), p. 5.

Within a very few months, Los Alamos forged a sense of community — and many of the wives credited Oppenheimer. Early on, in a nod to participatory democracy, he appointed a Town Council; later it became an elected body, and though it had no formal power, it met regularly and helped Oppenheimer keep in touch with the community's needs. Here the mundane complaints of life on the mesa — the quality of PX food, housing conditions and parking tickets — could be vented with gusto. By the end of 1943, Los Alamos had a low-power radio station that broadcasted a little news, community announcements and music, drawn in part from Oppenheimer's large collection of classical records. In small ways he made it known that he understood and appreciated the sacrifices everyone was making. Despite the lack of privacy, the Spartan conditions and the recurring shortages in water, milk and even electricity, he infected people with his own special sense of jocular élan. "Everyone in your house is quite mad," Oppenheimer told Bernice Brode one day, "You should get on fine together."[74] (The Brodes lived in an apartment above the Cyril Smiths and Edward Tellers.) When the local theatre group put on a production of *Arsenic and Old Lace*, the audience was stunned to see Oppenheimer, powdered white with flour and looking stiff as a corpse, carried on stage and laid out on the floor with the other victims of Joseph Kesselring's murder mystery. And when in the autumn of 1943 a young woman, the wife of a group leader, suddenly died of a mysterious paralysis — and the community feared a polio contagion — Oppenheimer was the first to visit the grieving husband.[75]

In September 1943, after a whirlwind courtship, Oppie's secretary, Priscilla Greene, married a chemist on the Hill, Robert Duffield. Oppenheimer was supposed to have given away the bride at a Santa Fe wedding in the stately adobe home of Dorothy McKibbin.[76] But at the last moment, General Groves called Oppenheimer away for a meeting in Cheyenne, Wyoming. Priscilla was nevertheless touched when Oppie insisted on having her drive with him to his

---

[74] Bernice Brode, *Tales of Los Alamos*, p. 39.

[75] Smith and Weiner, p. 265; Bernice Brode, *Tales of Los Alamos*, pp. 23, 72.

[76] Over the years, Dorothy McKibbin hosted more than thirty weddings in her home, including the wedding of Peter Oppenheimer. (Nancy C. Steeper, *Gatekeeper to Los Alamos: The Story of Dorothy Scarritt McKibbin*, p. 47 of draft manuscript.)

train so that he, a husband of less than three years, could share his wisdom on the travails of marriage.

Oppenheimer House, Los Alamos
Courtesy of the Los Alamos Historical Society

Oppie was the cook of the household. He was still partial to exotic hot dishes like *nasi goreng*, but one of his stock dinners included steak, fresh asparagus and potatoes prefaced by a gin sour or martini. On April 22, 1943, he hosted the first big party on the Hill to celebrate his 39th birthday. He plied his guests with the driest of dry martinis and gourmet food — though always on the scant side. "The alcohol hits you harder at 8,000 feet," recalled Dr. Louis Hempelmann, "so everybody, even the most sober people, like Rabi, were just feeling no pain at all. Everyone was dancing." Oppie danced the foxtrot, in his usual old-worldly style, holding his arm stiffly in front of him. Rabi amused everyone that night when he took out his comb, wrapped it in toilet paper and played it like a harmonica.[77]

Kitty refused to play the social role of a director's wife. "Kitty was strictly a blue jeans and Brooks Brothers shirt kind of gal," recalled one Los Alamos friend.[78] Initially, she worked part-time as a lab technician under the supervision of Dr. Hempelmann, whose job it was to study the health hazards of radiation. "She was awful bossy," he recalled.[79] Only occasionally did she invite old Berkeley friends over for dinner. But she never had the kind of open house parties expected of the director's wife. The Oppenheimers' next-door neighbors, however, liked to entertain. Deke and Martha Parsons held many of these events. Oppie encouraged everyone to work hard and play hard. "On Saturdays we raised whoopee," wrote Bernice Brode, "on Sundays we took trips, the rest of the week we worked."

---

[77] Dr. Louis Hempelmann interview by Sherwin, 10 August 1979, p. 29.
[78] Anne Wilson Marks interview by Kai Bird, 5 March 2002.
[79] Dr. Louis Hempelmann interview by Sherwin, 10 August 1979, 8, 24.

On Saturday evenings the Lodge was often packed with square-dancers, the men dressed in jeans, cowboy boots and colorful shirts, the women wearing long dresses bulging with petticoats. Not surprisingly, the resident bachelors hosted the rowdiest parties. These dorm parties were fueled by a concoction of half lab alcohol and half grapefruit juice mixed into a 32 gallon G.I. can and chilled with a chunk of smoking dry ice. One of the younger scientists, Mike Michnovicz, sometimes played his accordion while everyone danced.

Occasionally, some of the physicists gave piano and violin recitals. Oppenheimer dressed up for these Saturday evening affairs, wearing one of his

Bernice Brode and Jim Tuck at a Los Alamos dance
Courtesy of the Los Alamos Historical Society

finely tailored, tweedy suits. Invariably, he was the center of attraction. "If you were in a large hall," Dorothy McKibbin recalled, "the largest group of people would be hovering around what, if you could get your way through, would be Oppenheimer. He was great at a party and women simply loved him."[80] On one occasion someone threw a theme party: "Come as Your Suppressed Desire." Oppie came dressed in his ordinary suit, with a napkin draped over his arm — as if to imply that he wished merely to be a waiter. It was a pose no doubt designed to reflect a studied humility rather than any real inner longing for anonymity. As the scientific director of the most important project in the war, Oppenheimer was actually living his desire.

On Sundays many residents went for hikes or picnics in the nearby mountains, or rented the horses boarded at the Los Alamos Boys School's former stables. Oppenheimer rode his own horse, Chico, a beautiful fourteen-year-old chestnut, on a regular route from the east side of town west towards the mountain trails. Oppie could make Chico "single-foot" — trot by placing all four hooves down at different times — over the roughest trails. Along

[80] Bernice Brode, *Tales of Los Alamos*, pp. 23, 72; Dr. Louis Hempelmann interview by Sherwin, 10 August 1979, p. 30; Dorothy McKibbin interview by Jon Else, 10 December 1979, p. 22.

the way he greeted everyone he encountered with a wave of his mud-colored pork-pie hat and a passing remark. Kitty was also a "very good horseman, really European trained." Initially, she rode "Dixie," a full standard bred pacer who had once run the races in Albuquerque. Later she switched to a thoroughbred. An armed guard always accompanied them.[81]

Oppenheimer on Mt. Wilson

Oppenheimer's physical stamina atop a horse or hiking in the mountains invariably surprised his companions. "He always looked so frail," recalled Dr. Hempelmann. "He was always so painfully thin, of course, but he was amazingly strong." During the summer of 1944, he and Hempelmann rode together over the Sangre de Cristo Mountains to his "Perro Caliente" ranch. "It nearly killed me," said Hempelmann. "He was on his horse with the single-foot gait, perfectly comfortable, and my horse had to go into a hard trot to keep up with him. I think the first day we must have ridden 30 to 35 miles, and I was nearly dead."[82] Though rarely sick, Oppie suffered from smoker's cough, the result of a four or five pack a day habit. "I think he only picked up a pipe," said one of his secretaries, "as an interlude from the chain-smoking."[83] Given to spasms of uncontrolled bouts of coughing, his face would sometimes flush purple as he persisted in talking through his cough. Just as he made a ceremony of mixing his martinis, Oppie smoked his cigarettes with singular style. Where most men used their index finger to tap ashes off the end of their cigarettes, he had the peculiar mannerism of brushing the ash from the tip by using the end of his little finger. The habit had so calloused the tip of his finger that it appeared almost charred.[84]

---

[81] Dr. Louise Hempelmann interview by Sherwin, 10 August 1979, p. 10; Bernice Brode, *Tales of Los Alamos*, pp. 56, 88–93; Dorothy McKibbin interview by Jon Else, 10 December 1979, p. 20; John D. Wirth and Linda Harvey, *Aldrich, Los Alamos: The Ranch School Years, 1917–1943*, (University of New Mexico Press, Albuquerque, NM, 2003), p. 261.

[82] Dr. Louis Hempelmann interview by Sherwin, 10 August 1979, p. 22.

[83] Anne Wilson Marks interview by Kai Bird, 5 March 2002.

[84] Peer de Silva, unpublished manuscript, p. 1, courtesy of Gregg Herken.

Gradually, life on the mesa became comfortable, if not luxurious. Soldiers chopped firewood and stacked it for use in each apartment's kitchen and fireplace. The Army also collected the garbage and stoked the heating furnaces with coal. Every day the Army bussed in Pueblo Indian women from the nearby settlement of San Ildefonso to work as housekeepers. Dressed in deerskin-wrapped boots and colorful Pueblo shawls and wearing a bounty of turquoise and silver jewelry, the Pueblo women quickly became a familiar sight around town. Early each morning, after checking in with the Army's Maid Service Office near the town water tower, they could be seen trudging along the dirt roads toward their assigned Los Alamos household for half a day — which is why the residents began calling them their "half-days." The idea — endorsed by Oppenheimer and administered by the Army — was that such maid service would allow the wives of project scientists to work as secretaries, lab assistants, school teachers or "computing-machine operators" in the Tech Area. This in turn would help the Army to keep the population of Los Alamos to a minimum. Maid service was assigned largely on the basis of need, depending on the importance and hours of a housewife's job, the number of young children and on occasions of illness. Not always perfect, this bit of army socialism greatly eased life on the mesa and helped to turn the isolated laboratory into a fully employed, working community.[85]

* * *

While most Los Alamos spouses adapted to the stark climate, isolation and rhythms of the mesa, Kitty increasingly felt trapped. She wanted desperately what Los Alamos could give her husband — but as a bright woman who thought of herself as a biologist, she felt stymied professionally. After a year had passed, she told Dr. Hempelmann that she didn't think the blood counts she was doing for

Party at Los Alamos
Courtesy of the Los Alamos Historical Society

him as a lab technician meant anything — so she quit. She also felt isolated

---

[85] Bernice Brode, *Tales of Los Alamos*, pp. 28, 33, 51–52.

socially. If she was in the mood, she could be charming and warm with friends or strangers. But everyone sensed that there was an edge to this woman. Often, she seemed tense and unhappy. At Los Alamos social gatherings she could make small talk with people, but as one friend put it, "She wanted to make big talk."[86] Joseph Rotblat, a young Polish physicist, saw her occasionally at parties or in the Oppenheimer home for dinner. "She seemed to be very much aloof," Rotblat said, "a haughty person."[87]

Oppenheimer's secretary, Priscilla Greene Duffield had an ideal perch from which to observe Kitty. "She was a very intense, very intelligent, very vital kind of person," Duffield recalled. But she also thought Kitty was "very difficult to handle."[88] Pat Sherr, a neighbor and the wife of another physicist, felt overwhelmed by Kitty's meteoric personality. "She was outwardly very gay and exuded some warmth," recalled Sherr. "I later realized that it wasn't any real warmth for people, but it was part of her terrible need for attention, for affection."

Like Robert, Kitty tended to shower people with gifts. When Sherr complained one day about the kerosene stove in her cabin, Kitty gave her an old electric stove. "She would give me gifts and envelop me totally," Sherr said.[89] Other women found her abrupt manner verging on insulting. But so too did many men, even though Kitty seemed to prefer the company of men. "She's also one of the very few people I've heard men — and very nice men — call a bitch," recalled Duffield. But it was also clear to Duffield that her boss trusted Kitty and turned to her for advice about all manner of issues. "He would give her judgment as much weight as that of anyone whose advice he chose to ask," she said.[90] Kitty was the kind of wife who never hesitated to interrupt her husband, even though he, of course, was the kind of man who was always finishing other people's sentences for them. "It never seemed to bother him," recalled one close friend.[91]

Kitty intimidated nearly everyone else. The Los Alamos security officer, Captain Peer de Silva, spent much of his time observing the Director's wife. He thought her both highly attractive and dangerous, and later wrote

---

[86] Pat Sherr interview by Sherwin, 20 February 1979.

[87] Joseph Rotblat interview by Sherwin, 16 October 1989, p. 8.

[88] Peter Goodchild, *J. Robert Oppenheimer*, p. 127.

[89] Pat Sherr interview by Sherwin, 20 February 1979.

[90] Peter Goodchild, *J. Robert Oppenheimer*, p. 127.

[91] Dr. Louis Hempelmann interview by Sherwin, 10 August 1979, p. 18.

of her: "Vivacious, obviously proud of her sexiness, highly intelligent and articulate. Good looking in a pug-nosed kind of way, good figure. Tough woman... somewhat short, well and firmly built, well dressed. Very much the No. 1 wife in a society of many scientific wives."[92]

* * *

Early in 1945 Priscilla Duffield had a baby and Oppenheimer suddenly needed a new secretary. Groves offered him several seasoned secretaries, but Oppenheimer rejected each of them until one day he told Groves that he wanted Anne T. Wilson, a pretty blond, blue-eyed, twenty-year-old whom he had met in Groves's office. The daughter of a Navy officer and a neighbor of Groves's in the Cleveland Park neighborhood of Washington, DC, Anne Wilson had played tennis with Groves at the Army–Navy Club before he recruited her one day to work as one of his personal assistants. A blunt-speaking firecracker of a young woman, Wilson initially refused the job offer, telling Groves, "You're too ornery!"

Wilson had heard all about Oppenheimer when one day he came by Groves's office in Washington. "He stopped at my desk — which was right outside the general's door — and we made conversation," Wilson said. "I was just practically dumbstruck because here was this legendary character and part of his legend was that all women fell on their faces in front of him." Groves later told her that he had offered Oppenheimer any number of secretaries, but that he had turned them all down. "He looked at me," Groves related, "with those blue eyes and he says, 'I think I'd like to have Miss Wilson.'"[93]

Flattered, Wilson agreed to move out to Los Alamos. Before she went, however, Lt. Col. John Lansdale (Groves's counter-intelligence chief) approached her with an offer: he would pay her $200 a month if she sent him just one letter each month reporting on what she saw in Oppenheimer's office. Shocked, Wilson flatly refused to serve as his informant. "I told him," she later said, "Lansdale, I want you just to pretend you never even mentioned such a thing to me." She said Groves had assured her that once she moved out to Los Alamos, her loyalties were to be to Oppenheimer. But, perhaps not surprisingly, she learned after the war that Groves had ordered that she be covered by surveillance whenever she left Los Alamos — after working in his office he believed Annie Wilson knew too much to be left unwatched.

---

[92] Peer de Silva, unpublished manuscript, p. 3.
[93] Anne Wilson Marks interview with Bird, 5 March 2002.

Upon arriving in Los Alamos, Wilson learned that Oppenheimer was sick in bed with chicken pox, accompanied by a 104 degrees fever. "Our thin, ascetic Director," wrote the wife of another physicist, "looked like a fifteenth-century portrait of a saint with his fever-stricken eyes peering out from a face checkered with red patches and covered by a straggling beard."[94] One day while Oppie was lying in his sick bed, Pat Sherr came by with her four-year-old daughter Lizzie and explained that the camp pediatrician wanted the child exposed to chicken pox before her mother gave birth to a new baby. "It was a very amusing afternoon," Sherr recalled. "He was very sweet with her, very sweet. He looked a mess, he was really full of it and his head was shaved and he had it [chicken-pox scabs] all over his head. So we put Lizzie in his bed and Kitty kept saying, 'Touch him,' and Lizzie didn't like this thing at all. But he was very sweet with her and he'd say, 'You can touch me here, this isn't such a bad one and then he would rub his arm against her leg. It was ridiculous! She never got it from him."[95]

Soon after he had recovered, Wilson was invited over to the Oppenheimer home for drinks. Oppenheimer served her one, and then another, of his famous gin martinis, and as she was not yet acclimatized to the altitude, the powerful concoction quickly went to her head. Wilson remembered having to be escorted back to her room in the nurses' quarters.

Oppenheimer, Kitty, and their children

Annie Wilson was fascinated by her charismatic new boss and deeply admired him. But at twenty, she was not attracted romantically to Oppenheimer, a married man nearly twice her age in 1945. Still, Anne was a beautiful young woman, smart and sassy — and people began to talk on the Hill about the Director's new secretary. Several weeks after her arrival, Anne began receiving a single rose in a vase, delivered every three days from a florist in Santa Fe.

---

[94] Jane S. Wilson and Charlotte Serber (eds.), *Standing By and Making Do*, p. 50.
[95] Pat Sherr interview by Sherwin, 20 February 1979, p. 29.

The mysterious roses came without a card. "I was totally baffled, so I went around in my childlike way, saying, 'I got a secret lover. Who is sending all these gorgeous roses?' I never found out. But finally, one person said to me, 'There is only one person who would do that, and that's Robert.' Well, I said it's ridiculous."

Los Alamos was a small town and soon rumors began circulating that Oppenheimer was having an affair with Wilson. She said it never happened: "I have to tell you that I was too young to appreciate him. Maybe I thought a forty-year-old man was ancient." Inevitably, Kitty heard the rumors and one day she confronted Wilson and asked her pointblank if she had designs on Robert. Annie was thunderstruck. "She could not have misread my astonishment," Wilson recalled.[96]

In the years to come Annie got married, Kitty relaxed, and an enduring friendship developed. If Robert had been attracted to Annie, the anonymous single red rose was a subtle gesture not out of character. He was not the kind of man who initiated sexual conquests. As Wilson herself observed, women "gravitated" to Oppenheimer: "He really was a man of women," Wilson said. "I could see that and I heard plenty of that." But at the same time, the man himself was still painfully shy and even unworldly. "He was enormously empathetic," Wilson said. "This was, I think, the secret of his attraction for women. I mean it felt almost that he could read their minds — many women have said this to me. Women at Los Alamos who were pregnant could say, 'The only one who would understand was Robert.' He had a really almost saintly empathy for people."[97] And if he was attracted to other women, he still seemed devoted to his marriage. "They were terribly close," Hempelmann said of Kitty and Robert. "He would come home in the evenings whenever he could. I think she was proud of him, but I think she would have liked to have been more in the center of things."[98]

* * *

Inside the barbed wire, Kitty sometimes felt like she was living under a microscope. The Army commissary often had foods and goods only available on the outside with a ration card. The theatre showed two movies a week

---

[96] Anne Wilson Marks interview by Bird, 14 March 2002.
[97] Anne Wilson Marks interview by Bird, 5 March 2002.
[98] Dr. Louis Hempelmann, interview by Sherwin, 10 August 1979, p. 25.

for only 15 cents a show. Medical care was free. So many young couples had babies — some eighty births were recorded the first year and about ten a month thereafter — that the small seven-room hospital was labeled "RFD" for "rural free delivery." When General Groves complained about all the new babies, Oppenheimer wryly observed that the duties of a scientific director did not include birth control.[99] By then, Kitty was pregnant again. On December 7, 1944, she gave birth in the Los Alamos barracks hospital to a daughter, Katherine, whom they nicknamed "Tyke." A sign was posted over the crib, saying "Oppenheimer," and for several days people filed by to take a peek at the boss's baby girl.[100]

Four months later, Kitty announced she "just had to go home [to Philadelphia] to see her parents." Perhaps it was postpartum depression, or the excess of martinis in the Oppenheimer home, or the state of her marriage, but Kitty was on the verge of an emotional collapse. "Kitty had begun to break down, drinking a lot," recalled Pat Sherr. Kitty and Robert were also having problems with their two-year-old son. Like any toddler, Peter was a handful. And according to Sherr, Kitty "was very, very impatient with him." A trained psychologist, Sherr thought Kitty "had absolutely no intuitive understanding of the children."[101] Kitty had always been mercurial. Her sister-in-law, Jackie Oppenheimer, observed that Kitty "would go off on a shopping trip for days to Albuquerque or even to the West Coast and leave the children in the hands of the maid. They had one maid, a German one, and she was a regular tyrant." Upon her return, Kitty would bring an enormous present for Peter. "She must have felt so guilty and unhappy," said Jackie, "the poor woman."[102]

---

[99] By June 1944, one-fifth of all the married women in Los Alamos were pregnant. Charles Thorpe dissertation, p. 276; Jane S. Wilson and Charlotte Serber (eds.), *Standing By and Making Do*, p. 92; Robert Serber with Robert P. Crease, *Peace & War*, p. 83.

[100] Bernice Brode, *Tales of Los Alamos*, p. 22.

[101] Pat Sherr interview by Sherwin, 20 February 1979.

[102] Frank and Jackie Oppenheimer interview by Sherwin; Peter Goodchild, *Oppenheimer: Shatterer of Worlds*, p. 128.

CHAPTER THREE

Oppenheimer's Place in History

STANDING ON THE SHOULDERS OF GIANTS

Everet Beckner
Deputy Administrator for Defense Programs,
National Nuclear Security Administration, Department of Energy

The quotation, "standing on the shoulders of giants," comes from Sir Isaac Newton: "If I have seen further it is by standing on the shoulders of giants." To use this quotation as a theme for the work done during the Manhattan Project is something of a challenge. Because the first thing that I've asked myself is, considering the time — the 1600s — to whom was Newton referring?

Everet Beckner

We think back in time to the giants in science or philosophy, or whatever topic you want to think about, from the perspective that we have today. Think of how early Newton was in the development of science. It is very difficult to come up with more than two or three or four names that might have been on his mind — people conceivably like Galileo, or Copernicus, names that really are very early in the development of science.

His quotation, however, is one which I think is useful for us today. I want to introduce you to a few others before I get started, because once you get into a topic like this, you get kind of interested in some of the other quotations that are from scientists whom you know something about.

The one that I am going to refer to first is one that I will turn to a little bit later:

> *In science, the credit goes to the man who convinces the world, not to the man to whom the idea first occurs.*

Now that quotation is from Darwin and I will return to it because it has a lot to do with some of the views of Oppenheimer himself.

There are some others I want to note, for instance, Niels Bohr, about whom any physicist is aware. This quotation was attributed to him some time between 1945 and 1962:

> *Anybody who is not shocked by this subject has failed to understand it.*

Pretty good quotation!

Following that, I'd like to call your attention to this one by Wernher von Braun in 1973:

> *Basic research is what I'm doing, but I don't know what I'm doing!*

I think that sentiment fits really well here at Los Alamos.

And then I want to return to a quotation that most frequently is attributed to Oppenheimer, and that is the one attributed to him in 1947 at MIT:

> *The physicists have known sin, and this is a knowledge which they cannot lose.*

I think that is troubling, because of Oppenheimer's use of the word "sin." For those of us who are of European heritage, or certainly those who are of early American heritage, there is such a conservative ethic upon our lives, and as a result there are very serious implications for many of us when the word "sin" is used. I would interpret this quote differently, today, and I would turn to the quotation that I brought to your attention first, from Darwin, to explain it:

> *In science, the credit goes to the man who convinces the world, not to the man to whom the idea first occurs.*

I believe that Oppenheimer led the movement that proved that nuclear energy could result in a nuclear weapon, and that is what made the difference. When you hear Oppenheimer quoted, you almost think that in many ways, he might have been relieved if the bomb had not worked. But it was inevitable — *it was inevitable*. We all know that now, we know it from our deep understanding of nuclear physics and the work of scientific giants. We know it from our groping attempts to understand the origin of the universe, and the fact that much of that knowledge does now seem to fit together in ways that are becoming coherent.

Albert Einstein

So it was inevitable. We need to think about others who were involved and the role that they played. Finally, I want to return to the larger question of what this all means, not just in the context of science, but in the broader perspective of world history, when we talk about the "shoulders of giants."

So let's just think, for a few moments, about the other "giants" of the time. Early in physics, of course, the ones that stand out most are Newton himself and then I might add four more which could be called the "super-giants" of physics. They are Michael Farraday, James Clerk Maxwell, Niels Bohr, and Albert Einstein. For those who specialize in physics, it is easy to agree that those are among the greatest physicists who have ever lived. There may be others. There certainly are physicists today who will fall under that category when we look back in another hundred years. But from today's perspective, these are probably the greatest and the "giants" to which we will point when we think about planning the future.

Following that group, there are other physicists who have distinguished themselves in a more specialized field. When you talk about nuclear physics — and that obviously is our theme, here — the "giants" who come to mind are Werner Heisenberg, Ernest Rutherford, Ernest Lawrence, Hans Bethe, Erwin Schrödinger, and Enrico Fermi. Now, for those of you asking why I haven't gotten to a few other names already, I've got my next list. But I believe that these scientists affected physics more broadly than the next group that I will be talking about. Their impact was indeed in the areas which we think of as nuclear physics, but it was somewhat broader than that, and furthermore was an impact which is undeniable. You won't find anyone who will argue with that when you use those names. Of that group, only one is still living, Hans Bethe.

In the next group, one certainly will encounter Oppenheimer — and Paul Dirac, Arthur Compton, Edward Teller, Richard Feynman, Eugene Wigner, Steven Weinberg and John von Neumann. I could easily have added more names to that list, but I tried to keep it somewhat short! These are the scientists whom I believe were responsible for application of this understanding that evolved from the earlier giants of nuclear physics. They contributed in many ways, but particularly because they really understood how to apply their ideas. Out of this work came the nuclear weapons program. Out of this work came the nuclear reactor program. This group of physicists was responsible for both the peaceful and the military uses of the nuclear energy and all that flowed from that work.

Enrico Fermi

And this development really led us to "big science," as we think of it today. You look around today and you will find that scientific journal articles are as likely to have forty authors as they are to have four, let alone one. Science has changed greatly during this time, as has the world. It did so riding on the shoulders of these giants.

Now where have we gone since then? All of this work, for the most part, was done before 1950, certainly before 1960, and it is now forty years later. And what we have is a whole new set of giants. They are working in other areas that are offshoots of earlier work in physics. Probably the field that first comes to mind is what we generally call "microelectronics." That has come out of the work of early physicists who were looking at other applications for the same ideas. The eminent scientists in this field are John Bardeen, Arthur Schawlow and Charles Townes who were working on early ideas of transistors and lasers. You have another set of people working on nuclear fusion and the names that really come first are Bethe and Teller. Out of that, more broadly, people have come to identify cosmology as a legitimate science — in fact, it is the dominant science now, as we look to the future. And there are giants in this field that are only just being identified. Many are yet to be fully identified.

The next set of scientists I'd like to mention are in biology — another area of study that will change the world. And in that regard, it is easy to bring it all back to Darwin again. Darwin clearly set the stage for what's going on now in genetics and cellular biology, and the whole field is exploding. From this line of research we are going to see things like genetic manipulation and artificial intelligence, which will be a combination of all those microelectronics, lasers, and cellular biology. Somewhere in there we are going to figure out how to improve upon the brain.

And a few people are beginning to occupy that stage. The name that I would think of first is Steven Pinker. So you do pick up these names, and you see some of the giants developing, and you recognize that they all have in fact stood on those shoulders of the earlier giants.

So, that is the way it works for all of us. I think that our future work depends greatly upon the people who have helped us get started. If they turned out to be particularly illustrious people, then obviously we got an early chance to stand on some large shoulders. And many of the people in this audience, I think, can identify people who were that important to them, who have made a real difference in their work.

Let me now put on my Washington hat and say just a few words about the larger scene as I see it today. The question is how do we move the world forward, in circumstances which look extraordinarily intractable, more so than I've seen in many years — and I've now been around long enough to have seen quite a few difficult times. In order to approach this question, we can look back at history again, at some of the giants, but in this case they are of a different kind.

I happened to have just finished reading a book on the life of Genghis Khan, and indeed, there's a lot to be learned from him as we view the troubles of the world today. He was able to conquer half the world, half the civilized world of the time — or at least he, and his sons and his grandsons did this during the 13th and 14th centuries. From China to the Mediterranean, from southern Russia to India, Genghis Khan did it by coming up with a new way of waging war, as it turned out. He gave every one of his soldiers a horse, and went to war. And pretty soon he won, every time. And he ruled ruthlessly. Thereby Genghis Khan and his descendents controlled that world for more than one hundred years.

Now, this domination came to an end for a very strange reason. I do not think people realize this, but the black plague essentially was the transitioning event because it forced all communication, transportation and interactions to cease. As everyone around them was dying, people realized that the only way to keep from getting the black plague was not to interact with anybody else. So they shut themselves up in their houses. They stopped moving. They stopped commerce. Of course, huge numbers of them simply died. But by the time the black plague subsided, the empire of Genghis Khan had more or less disintegrated and had become disconnected from the rest of the world.

One thing had happened, however, to set the stage for the next great events of the world. Kublai Khan, Genghis Khan's grandson, had tried twice to invade and conquer Japan by sea, using ships that were built in China. He failed both times. But that was the beginning of sea warfare. Kublai Khan's army did not do very well at it, but the giants who came along next, such as Henry the Navigator and obviously Christopher Columbus, did. The nations which turned their armies into sea warriors became the great sea powers: Spain, France, and England. That was the beginning of sea power which then led to a whole new process for waging war and controlling the world.

That lasted more than a hundred years. The next set of giants began to realize that there was yet another way to control political situations, which turned out to be the airplane. The Wright brothers and other pioneers of flight led to a fundamental shift in the way we interact with each other on the world stage, and the way people wage war.

One of the next steps in this progression was the development of nuclear weapons, which was the next defining event in world politics as well as in science. Along the way, too, there was the development of nuclear submarines by scientists and engineers. These strategies involving nuclear weapons played out reasonably well in the 50s and 60s, and 70s and 80s, under a strategy which we ended up calling "mutually assured destruction." We had too many nuclear weapons for the Russians to ever think about attacking us, and they had too many nuclear weapons for anyone else in the world to think about attacking them. So we lived together in something that was very tenuous — a peace with trouble from time to time — for the better part of forty years.

Now, we've reached today. It is clear we don't know how to use this power to control the world at this point in time. We don't know how to use the weapons we have to control the world and its terrorism. The last guy who knew how to do that, in fact, was Genghis Khan! Now, that is not a very pleasant analogy to use for thinking about the future. After all, Genghis Khan did it just by killing everybody who didn't agree with him. When he would come to a town, he would first send in messengers telling them that if they would all agree to surrender, he would come in and take over. If they wouldn't, he was going to kill them. And that is exactly what he did.

I do not think any of us believe that to be the way to go forward at this point in time. But the problem is we do not know how to go forward. And it is with this thought that I am going to close. It is the interplay of science and world politics which has provided the great policy and technology problems that have driven the course of history for the past six hundred years. Recently, for instance, the chemists helped us find ways to make poison gas, which was used in the First World War. The physicists came up with a way to make nuclear weapons, used in the Second World War, and as deterrents to war for the next fifty years. Right now, I don't think we have a solution to the terrorism problems of the world. So, I think we need first to recognize that fact, and then find ways to craft the solutions. That is the troubling world we live in, but it is the real one. It may be that, at this particular point in time, we need the shoulders of a few political giants, rather than scientific giants, to stand upon.

THE CAUTIONARY TALE OF ROBERT OPPENHEIMER

Gregg Herken
Professor, School of Social Sciences, Humanities and Arts,
University of California, Merced

Gregg Herken

The notion that scientists have no "proprietary rights" to say what should be done with their inventions was expressed by Robert Oppenheimer a month before the explosion of the first atomic bomb. A month after two atomic bombs had been dropped on Japan, not even Oppenheimer believed in this prescription. In 1949, Oppenheimer would oppose development of the hydrogen bomb on both practical and ethical grounds. The loyalty hearing that took place five years later suggests that science in the service of the state bears a potential cost, for both sides.

Some fifty years ago, when asked what impact the Cold War had had upon his discipline, Princeton physicist Henry DeWolf Smyth reportedly replied: "Secrecy, scientists who will take orders, and big equipment." Two of the things that Smyth mentioned — secrecy and big equipment — are undoubted legacies of the Cold War. But the third — the willingness of scientists to take orders — is a more problematic and complicated tale, precisely because of the Cold War, and because of the particular experience of one scientist who answered the summons: Robert Oppenheimer.

When he was chosen by General Groves to direct the Los Alamos laboratory in the fall of 1942, Oppenheimer showed no hesitation about working on the atomic bomb. Indeed, after Oppie came under suspicion by Army counter-intelligence for his prewar left-wing views, it was Oppenheimer's ambition, not his patriotism, which convinced Groves's head of security, John Lansdale, that Oppie would not and could not be a spy for the Russians.

Near the end of the war, when Oppenheimer was chosen to serve on a scientific panel (with Enrico Fermi, Ernest Lawrence, and Arthur Compton) to advise on the use of the bomb, it was Oppenheimer, the head of the panel, who argued that there was no practical alternative to military use of the weapon against Japan. But as Oppenheimer also wrote to Secretary of War Henry Stimson at that time:

> With regard to these general aspects of the use of atomic energy, it is clear that we, as scientific men, have no proprietary rights. It is true that we are among the few citizens who have had occasion to give thoughtful consideration to these problems during the past few years. We have, however, no claim to special competence in solving the political, social, and military problems which are presented by the advent of atomic power.

Even at the time that Oppenheimer wrote those words, his was probably already a minority view among scientists working on the project. Leo Szilard obviously felt no such compunction about telling the government what it should do when he passed around his petition at the University of Chicago. Edward Teller claims that he was sympathetic to Szilard's appeal but that Oppenheimer (who reminded Teller that it was not the job of scientists to decide how the bomb was used) forbade him from circulating the petition at Los Alamos, as Szilard had requested.

Within weeks of the bombing of Hiroshima and Nagasaki, Oppenheimer himself had reversed his stand on the role of scientists. By September 1945, Oppie was using his status as the "father" of the atomic bomb to importune top officials in the Truman administration and even Truman himself on behalf of his newfound cause: the international control of atomic energy.

Arguably the most famous, or notorious, instance of scientists speaking the truth to power came some fours years later, in October 1949, when Oppenheimer, then chairman of the Atomic Energy Commission's General Advisory Committee, drafted the GAC's majority report, urging that the nation *not* proceed with a crash effort to develop the hydrogen superbomb — which the committee described as potentially "a weapon of genocide."

As we know, the GAC's advice was ignored, and a few years later (after the prototype superbomb had been successfully tested) Oppenheimer was brought to account for his views. Although the impetus behind the revocation of Oppenheimer's security clearance and the resulting loyalty hearing concerned lies that Oppie had told to Army security agents during the war, the evidence now available in AEC and FBI files leaves little doubt that the real motivation behind the Oppenheimer trial (and its verdict) was Oppie's so-called "failure to enthuse" over the hydrogen bomb. The intent, therefore, was not only to end Oppenheimer's influence as a science adviser — which was essentially at an end by this time anyway — but to "unfrock" Oppenheimer before his peers.

Oppenheimer at Princeton

With one dissenting vote, the AEC declared Oppenheimer a loyal citizen, but stripped him of his security clearance, just one day before it was due to expire. The verdict had two effects: it made Oppenheimer a martyr in the scientific community, and it was the start of Oppie's academic exile at Princeton, which would endure for nearly a decade. The message that the state seemed to be sending to its scientists in the Oppenheimer case was, "we value your necessary inventions, but not your unwanted advice."

Oppenheimer would be partially rehabilitated in 1963, when he was awarded the Fermi medal, but the Kennedy administration pointedly refused to reinstate his security clearance on that occasion, lest the old controversy be revived. Although Oppie died in 1967, his ghost is still very much with us — as Edward Teller reminded in his memoirs, published just last year. Forty years after the infamous loyalty hearing, Teller would attribute the difficulty he had in recruiting scientists to work on Ronald Reagan's Strategic Defense Initiative to the lingering aftermath of the Oppenheimer case.

Today, nearly sixty years after the Oppenheimer trial, secrecy is with us more than ever; big equipment as well. But the era when scientists were blindly willing to follow orders — even in wartime — may well be over.

THE EARLY YEARS OF ROBERT OPPENHEIMER

Jon Hunner
Associate Professor, Director of Public History Program at New Mexico State University

Excerpt from *Chasing Oppie*, a forthcoming book from the University of Oklahoma Press.

At the beginning of the twentieth century, a boy was born in New York City who would help make it the *American* Century. As an adult, his research in nuclear physics, his work on the Manhattan Project, and his advocacy for civilian control of atomic weapons, first helped create the Atomic Age and then to direct it. Robert Oppenheimer was a complex person and among the various perspectives about him that will be offered at this conference, I would like to add this. As a young man, the American West greatly impacted Oppenheimer. And in return, Oppenheimer transformed the West. But in the beginning, Robert Oppenheimer was a New Yorker.

Jon Hunner
Photo by Darren Phillips

His parents, Ella and Julius Oppenheimer, lived in a fashionable part of New York, on the upper west side of Manhattan. Ella's family, the Friedmans, was of European descent and had been in America for several generations. As a painter, she had studied in Paris, and then returned to New York to teach art. She always wore a curious glove on her right hand which elicited stony silences when noted by visitors. She was born with a disabled hand, so the glove contained an early prosthetic device that allowed movement of an artificial thumb and forefinger.

Julius had come to America more recently than Ella's family, having arrived in 1888 when he was only seventeen. He had left Hanau, Germany, where his father was a farmer and grain merchant. As soon as Julius got off the passenger ship at New York, he started working in the city for two uncles in their textile import business. Julius quickly worked his way up through the ranks and became a successful businessman with sophisticated tastes.

Robert and Frank Oppenheimer

Ella and Julius possibly met at an art exhibition in New York. They married in 1903, and on April 22, 1904, Robert Oppenheimer was born to the couple at their apartment on West 94th Street after a long labor. His birth certificate reads "Julius Robert Oppenheimer," which in later years was shortened to "J. Robert Oppenheimer." Although both parents were of Jewish descent, they did not strictly practise their religion. In fact, the Ethical Cultural Movement, a secular off-shoot of Judaism, attracted both parents. Founded in 1876 by Felix Adler, the Ethical Cultural Movement sought to create a "Judaism of the Future" based on deeds of beneficial social activities, in particular the moral and intellectual education of the working classes.[103] The Workingman's School that Adler created became so successful in attracting working class students and in providing a good education that upper class Jews like the Oppenheimers (whose children were barred admittance to many private schools at the time because of anti-semitism) sought to have the school's doors open to their children. By the time that Julius served on their Board of Trustees from 1907 to 1915, the renamed Ethical Cultural School had expanded to accept such children so that Robert entered it in 1911.[104]

---

[103] Jeremy Bernstein, *Oppenheimer: Portrait of an Enigma*, p. 9.

[104] Bernstein, p. 6; Gregg Herken, *Brotherhood of the Bomb*, p. 12; Peter Goodchild, *J. Robert Oppenheimer: The Shatterer of Worlds*, pp. 10–11; Schweber, pp. 42–43, 48–49.

Central to the mission of the Ethical Culture movement was the belief that "man must assume responsibility for the direction of his life and destiny." Humans must answer to themselves, not to God, for their actions. For a person who will create a weapon that killed tens of thousands of people and which could destroy life as we know it, this belief would help explain Oppenheimer's actions after the atomic bombings of Hiroshima and Nagasaki.[105]

Later in life, Oppenheimer remembered that when he was ten or twelve, his main interests were "minerals, writing poetry and reading, and building with blocks."[106] On one of the family's trips to Germany when Robert was young, a grandfather, Ben, gave him a small collection of rocks with labels written in German. The rocks fueled a scientific interest because he tried to understand what he saw in them, things like the structure of the crystals, the polarized light, and how rocks and crystals were formed. Oppenheimer corresponded with members of the New York Mineralogy Society and eventually was invited to join and then give a paper at one of their meetings. Accompanied by his father, they showed up at the meeting place only to be told that the Robert Oppenheimer they expected to appear was not a boy of eleven. The next youngest member of the Society was in his seventies.[107]

Oppenheimer's parents fostered his brilliance. Later in life, Robert recalled: "I think that both [my father] and my mother were pleased that I was a good student, were pleased that I was highbrow, were perhaps somewhat mockingly proud of my vigor in collecting and learning about minerals. . ."[108] His intellectual intensity did worry his mother though: "I think my mother especially was dissatisfied with the limited interest I had in play and in people of my own age, and [. . .] I know she kept trying to get me to be more like other boys, but with indifferent success."[109]

On August 14, 1912, Robert gained a younger brother when Frank Oppenheimer was born. Another brother had died in infancy before Frank

---

[105]S. S. Schweber, *In the Shadow of the Bomb*, pp. 50, 52–53; Box 294, "Printed Matter," letter from Hillman to Hagy, p. 1, Box 294, J. Robert Oppenheimer papers (JRO), Library of Congress, Manuscript Division (LC-MD); Richard Rhodes, *The Making of the Atomic Bomb*, p. 119; Bernstein, p. 10.

[106]Schweber, p. 53.

[107]Alice Kimball Smith and Charles Weiner, *Robert Oppenheimer: Letters and Recollections*, p. 3; Goodchild, p. 11.

[108]Smith and Weiner, p. 5.

[109]Smith and Weiner, p. 5.

was born. Eight years Frank's elder, Robert was both a big brother and, as they became adults, a mentor. Perhaps not quite as brilliant as Robert, Frank still held his own around the family's dinner table, and throughout most of their lives, the brothers were close. They shared many common interests, from their passion for the West, to their pursuit of physics, to their involvement in left-wing causes. Their support in the 1930s of liberal organizations in California came back to first haunt and then destroy both brothers' lives in the 1950s.

Robert had already begun attending the Ethical Cultural School the year before Frank's birth. At the school where brilliance was expected, Robert shone. His Greek and Latin instructor, Alberta Newton, recalled that "he received every new idea as perfectly beautiful." Even at an early age, Robert was quick to grasp complex concepts and ideas and impatient in waiting for his peers to catch up to him. So, he was often sent to the library to do advanced work in math and when he returned, he explained it to the other students.

As a teenager at the Ethical Culture School, Robert was thin and gangly, with a mass of brown hair and vivid blue eyes. Two teachers at the school had a great impact on Oppenheimer: Augustus Klock and Herbert Smith. Klock taught chemistry and physics to the school's high school students. Building upon the Oppenheimer's fascination with mineralogy, Klock sparked a deeper interest in science and the mechanics of our physical world. Klock introduced Robert to the mysteries of the universe at the laboratory tables of the class-room, and for a person with an inquiring mind and an active imagination, the connection between what he saw in experiments and how that answered questions both practical and profound satisfied the young man's hunger for knowledge.[110]

The other teacher that greatly influenced Oppenheimer was Herbert Smith. Smith recalled: "Robert simply towered above all his brilliant contemporaries. He is certainly the most brilliant man I have ever met. . . It was immediately obvious that he was a genius."[111] In Smith's English class, which was a college preparatory course, Robert brought his already lively interest in literature and poetry. Even though he usually sat and listened, Smith remembered him as "a flawless student."[112] For years after Oppenheimer had left the Ethical Culture School, he still exchanged poems and letters with Smith.

---

[110]Smith and Weiner, p. 4.

[111]"Printed Matter," Letter from Hillman to Hagy, p. 3, Box 294, JRO, LC-MD.

[112]"Printed Matter," Letter from Hillman to Hagy, p. 4, Box 294, JRO, LC-MD; Smith and Weiner, p. 5.

Part of the troubles that Robert confronted was his awkwardness in social settings. When he was fourteen, he attended a summer camp. In the midst of an energetic outdoor life, his brainy nature and know-it-all attitude set him up to be tormented by the other boys. A letter home that said that he was learning about the "facts of life" resulted in a crackdown on the boys telling dirty jokes and stories. As punishment, the other boys locked Robert overnight in an icehouse without any clothes.[113]

A different view of the young Oppenheimer emerges from a classmate of his. Jane Didisheim still held sharp memories of Robert fifty years after she went to the Ethical Culture School with him: "He was very frail, very pink-cheeked, very shy, and very brilliant of course. Very quickly everybody admitted he was different from all the others and very superior. . . Aside from that he was physically [. . .] rather undeveloped, not in the way he behaved but the way he went about, the way he walked, the way he sat. There was something strangely childish about him."[114]

In addition to his easy mastery of math, science, and literature, Robert also learned languages quickly. He wrote sonnets in French, spoke German, picked up some Chinese. His quick mastery of new languages continued after he left the school. In the 1930s, he taught himself Sanskrit so that he could read the sacred texts of Hinduism, and once when he was a visiting professor at the University of Leiden in the Netherlands, he learned Dutch in six weeks so he could present a lecture in that tongue. Smith reacted to Robert's intellect the way that many did: "His mind [was] so tremendous that it makes you really uneasy."[115]

One of the schoolmates that felt at ease with Robert was Francis Fergusson who came to New York from the desert Southwest. To prepare him for Harvard University, Francis left his native Albuquerque, New Mexico, and attended the Ethical Culture School. There, he met and became good friends with Robert. Francis went on to write plays, essays and histories of the arts, produce theatrical pieces, and became a fixture in the arts and literary sphere of New Mexico. For Robert, Francis not only held a similar keen intellect and intensity, but he also offered a window into a different way of life, a Western sensibility of vast panoramas, ancient cultures, and wide open opportunities.

---

[113] Smith and Weiner, p. 6.

[114] Smith and Weiner, pp. 6–7.

[115] "Printed Matter," Letter from Hillman to Hagy, pp. 5–7, Box 294, JRP, LC-MD.

From their senior year at the Ethical Culture School, through their undergraduate and graduate schooling, the New Yorker and the Westerner were the best of friends.

Robert graduated from the Ethical Culture School in February 1921 at the age of seventeen. That summer, he traveled to Europe with his family. Robert and Frank went on a prospecting trip to Bohemia, where he contracted a severe case of dysentery. Already frail, the sickness forced him to postpone his entrance into Harvard as he recovered, first in Europe and then back in the States. Along with the dysentery, he also was struck with colitis. The extended convalesce, first at his parents' apartment in New York and then with several trips, not only allowed him to recover but also to experience new parts of the country. In the spring of 1922, he traveled to the South. By that summer, Robert had recovered enough to plan an extended trip to the Southwest. To cap off the year of recuperation with a vigorous trip west, Robert's parents asked the genial Herbert Smith to accompany their son in his travels.

The Southwest offered dramatic landscapes with high mountains and distant vistas, peoples from the unique cultures, and a world view distinctly different from anything even the travel-savvy Robert had experienced in Europe. Going West also offered an escape from the ills not only of one's body, but of industrial America. At about the same time that Robert and Herbert disembarked from the train in New Mexico, other people from the east — intellectuals, writers, artists — sought refuge from the disappointments of the failed peace after World War I and from the rampant consumerism and hustling industrialism of the Roaring Twenties. Some of them found that refuge in the mountains and towns of New Mexico.

For weeks, Robert and Herbert hiked and rode horses through the steep mountains, camped outdoors, and stayed at guest ranches. Their headquarters was Los Piños, a guest ranch at Cowles, high in the Sangre de Cristo Mountains. Los Pinos was run by Katherine Chaves Page and her new husband Winthrop. Katherine came from an established Hispanic family headed by don Amado Chaves, and they accepted Oppenheimer and Smith into their circle of family and friends. Smith later commented that because of Robert's acceptance by the patriarch don Amado and the aristocratic Chaves clan, "for the first time in his life, [Robert] found himself loved, admired, sought after. . ."[116] For an insecure and sheltered young man, this acceptance did as much as the crisp desert climate to cure him of his ailments.

Another family that embraced Robert in New Mexico was the Fergussons. He and Herbert visited them in Albuquerque where they met Francis's brother Harvey and sister Erna. Both Harvey, as a novelist, and Erna, as a Southwest writer, would distinguish themselves in the coming years. The Fergussons also introduced Robert to Paul Horgan, a dashing young New Mexican who also became a famous writer. Once Robert left New Mexico, he wrote

Sangre de Cristo Mountains
Courtesy of the National Archives and Records Administration

to many of his new friends and stayed connected to the desert Southwest through correspondence and rendezvous with them on the East Coast. For a young man who had traveled throughout Europe, the people Robert met in provincial New Mexico that summer of 1922 remained friends for years. Their companionship, liveliness, and alternative lifestyles counteracted Robert's conventional East Coast upbringing.

When one is exposed to new cultures, a code switching occurs. In linguistics, code switching happens when a person changes from his dominate language to a different one. For example, in New Mexico, one often hears Spanish and English spoken in the same sentence, *que no?* This linguistic code switching allows people to express themselves better by choosing words or expressions that their native tongue does not adequately communicate. Code switching also shows that a person is sophisticated, that he or she can speak in more than one language. A cultural code switching also occurs when a person finds something in another culture that is then incorporated into one's own beliefs and behavior. Through the centuries, clothing styles, cuisine, music, religion, and other cultural attitudes have been freely borrowed whenever two

---

[116]Smith and Weiner, p. 10.

or more cultures come together. In the mountains of New Mexico, Robert interacted with different cultures and adopted some of their beliefs and cultural resources, which helped form his character.

Another part of their trip to New Mexico introduced Robert to a place that would dramatically alter his future. Robert, Herbert, and several other friends rode horses through Frijoles Canyon, where Native American ruins from the 1300s hid among the stunted piñon pine and tall ponderosa pine trees. Riding through the deeply crevassed plateaus and mountainsides, they encountered an isolated boys' school called "Los Alamos." Twenty-one years later, Robert would pick Los Alamos as the place to create the atomic bomb.

The time in the fresh air with vigorous physical activity challenged Robert's frail frame, and he prospered. The trip to the Southwest added the final touch to his recovery, and he impressed Herbert with his stamina and natural ability with horses. In fact, Herbert commented that Robert displayed a fatalistic attitude toward physical danger that bordered on the reckless. Robert would return to the high mountains of New Mexico time and time again over the next two decades to recover from his hectic life. He would visit it, write about it in letters to friends, and eventually would buy land there before he even bought a home of his own. And from New Mexico, he would launch a new age that changed the world.[117]

Back on the East Coast, one of the places where the Oppenheimer family entertained their friends was at Bay Shore, a town on the Atlantic Ocean side of Long Island, about fifty miles away from Manhattan. On an estate of six acres, the family spent many summers away from the heat of the city. Sailing one of their two boats occupied their time. *Lorelei*, a 40-foot yacht, and a 28-foot sloop, were tied up at the pier on the property. Robert and Frank named the sloop *Trimethy* after the colorless liquid trimethylamine, which smelled like pickled herring. The brothers took *Trimethy* and any visiting friends out on the bay between their house and the outer island. Not known for physical prowess, Robert nonetheless was an aggressive, even at times a reckless sailor. One time, Robert and Paul Horgan sailed too close to the Fire Island inlet and were carried out on the tide into the storm-tossed Atlantic breakers. For several hours, they struggled to tack back into the more tranquil bay. Once back in the bay, their progress back to the house was slow. Robert's worried father sent

---

[117] Rhodes, p. 121; Smith and Weiner, p. 10.

a revenue cutter out to search for them and around 11 p.m., it rescued them. When they returned, however, Robert's parents did not admonish them. And the mad dashes across the bay on *Trimethy* continued.[118]

From Horgan's visits to the Oppenheimers, both in New York and at their summer house on Long Island, he remembered: "There were high spirited goings on all the time. I think it is perfectly right to say that even then — and all my life I've felt this — he was the most intelligent man I've ever known, the most brilliantly endowed intellectually. And with this, in that period of his life, he combined incredibly good wit and gaiety and high spirits."[119]

Another aspect of Oppenheimer's personality emerged as he grew. He at times suffered from depression. Horgan recalled that: "Robert had bouts of melancholy, deep, deep, depressions as a youngster... He would seem to be *incommunicado* emotionally for a day or two at a time." Others also observed a moodiness in the young man, which combined with his social fumbling, arrogance, and lack of patience, shows that he was troubled at times. In 1923, he wrote to Francis Fergusson: "I find these awful people in me from time to time, and their expulsion is the sole excuse for my writing." Robert dealt with this depression off and on throughout his life.[120]

After taking a year off to regain his physical as well as his mental health, Robert entered Harvard University in the fall of 1922. He planned on completing the four-year degree in only three. To do this, he pushed himself hard. He arrived at the laboratories early, took heavy loads of classes, and still had time to audit courses in subjects not required for his degree. For most students, a normal semester load of courses at Harvard was five, but Robert took six courses for credit and often audited more. In one year, he took for credit four chemistry classes, two in French literature, two in mathematics, one in philosophy, and three more in physics. A Harvard classmate noted that Robert "intellectually looted the place."[121]

In a letter to Smith that first winter at Harvard, Robert pined for the open spaces of the West. Replying to Smith's plans to summer in New Mexico, Robert wrote: "Of course, I am insanely jealous. I see you riding down from the mountains to the desert at that hour when thunderstorms and sunsets

[118]Smith and Weiner, pp. 34, 37; Goodchild, p. 34.
[119]Smith and Weiner, p. 8.
[120]Smith and Weiner, pp. 9, 57.
[121]Rhodes, p. 121.

caparison the sky; I see you in the Pecos [. . .] spending the moonlight on Grass Mountain; I see you vending the marvels of the upper Loch, of the upper amphitheater at Ouray, of the waterfall at Telluride, the Punch Bowl at San Ysidro — even the prairies around Antonito — to Philistine eyes."[122] For many people, a summer sailing on Long Island Sound and lounging at the family estate waited on by servants would be ideal. For Robert, he longed for the landscape, the people, and the freedom of the West. Despite several plans to avoid the summer on the Long Island estate, first through a feigned recurrence of illness and then by trying to persuade his parents to accompany him to New Mexico, Robert spent the summer of 1923 on the East Coast. He admitted to Smith in a letter: "I fear that if I transported father and mother to the midst of the desert and dropped them, I should jeopardize my puny inheritance and to chaperone them to Los Piños would insure a new nervous breakdown."[123] Although family and studies kept Robert in the East once he entered Harvard, he often thought about New Mexico and corresponded with mutual acquaintances about their lives and adventures there.

Because of his heavy load of courses, his outside reading, and self-imposed haste to graduate in three years, Robert had only limited time for a social life. His roommate, William Boyd, mentioned that he never saw Robert out on a date. Another friend, Jeffries Wyman, admits that none of them had much time for dating: "We were all too much in love with the problems of philosophy and science and the arts and general intellectual life to be thinking about girls." For Wyman, Robert's unease with people was also a hindrance: "He found social adjustment very difficult, and I think he was often very unhappy... I suppose he was lonely and felt he didn't fit in well with the human environment... We were young people falling in love with ideas right and left and interested in people who gave us ideas, but there wasn't the warmth of human companionship perhaps." For Robert and his cohort of friends at Harvard, ideas and intellectual challenges excited them. They were drunk on ideas and in love with discussing them.[124]

The professor who taught the thermodynamics course at Harvard, Percy Bridgman, was a distinguished experimental physicist who changed Robert's life. Robert later said: "I can't recall how it came over me that what I liked in

---

[122] Smith and Weiner, pp. 22–23.
[123] Smith and Weiner, p. 35.
[124] Smith and Weiner, pp. 60–61.

chemistry was very close to physics."[125] For him, physics went to the heart of matter, even more than chemistry did: "It was the study of order, of regularity, of what makes matter harmonious and what makes it work."[126] Physics appealed to the philosopher in Oppenheimer more than chemistry and even though he did not attend any courses in physics his freshman year, he petitioned the physics department to allow him to start taking graduate level courses. He was permitted to jump over the more basic physics' courses, with one of the profes-

Oppenheimer at Harvard

sors commenting: "Obviously, if he says he's read these books he's a liar, but he should get a Ph.D. for knowing their titles." Despite his attraction to physics, in June 1925, Oppenheimer graduated *summa cum laude* with a major in chemistry. Under his yearbook picture, where a less modest person might have bragged about his achievements at one of the most prestigious universities in the land, Robert simply wrote, "In college three years as an undergraduate."[127]

After Harvard, Oppenheimer wanted to attend the Cavendish Laboratory at Cambridge University in England. He later admitted: "I don't even know why I left Harvard, but I somehow felt that [Cambridge] was more near the center." With Nobel prize winner Sir Ernest Rutherford directing Cavendish, Cambridge was perhaps the center of experimental physics. Professor Bridgman wrote a letter of recommendation to Rutherford highlighting Oppenheimer's

> perfectly prodigious power of assimilation. . . His weakness is on the experimental side. His type of mind is analytical, rather than physical, and he is not at home in the manipulations of the laboratory. . . It appears to me that it is a bit of a gamble as to whether Oppenheimer will ever make any real contributions of an important character, but if he does make good at all, I believe he will be a very unusual success. . .

[125] Smith and Weiner, p. 45.
[126] Goodchild, p. 16.
[127] Smith and Weiner, pp. 28–30, 39; Goodchild, p. 16.

Rutherford was not impressed with Bridgman's qualified endorsement and did not accept him as one of his students; however, Rutherford did arrange for Oppenheimer to work with another experimental physicist in his laboratory, Nobel prize winner, J.J. Thomson.[128]

With his entrance to Cambridge finalized, Robert went to the mountains of New Mexico before beginning the arduous work of pursuing a Ph.D. in physics. This was his first time west since he had roamed the forested high

Sangre de Cristo Mountains

desert plateaus with Herbert Smith in 1922. Although he had not visited for three years, he had kept in touch with the friends he had made in New Mexico. On one occasion, he sent at considerable expense a cake made in New York City for the 70th birthday celebration of Amado Chaves, the patriarch of the family that had adopted Oppenheimer and Smith in 1922. On the cake was the Chaves family crest. For this visit to the West, Robert had the rest of his family with him. Frank stayed with him at a ranch near Cowles, and their parents resided in the more luxurious Bishop's Lodge in Santa Fe with side trips up to the mountain retreat.

The visitors enjoyed a wide range of activities: trips to Native American dances, carousing at the annual Santa Fe Fiesta, but most of all, riding horses up to the high peaks that towered over the cabin at Cowles. Paul Horgan visited the group and described one of their jaunts. They hired horses in Santa Fe (which is 7,000 feet above sea level) to ride east over the Sangre de Cristo Mountains and down into Cowles. As the crow flies, the distance is about fifteen miles, but the peaks that they ascended rose to well over 10,000 feet above sea level. As Horgan recalled: "It turned out to be a day-long venture, full of merriment and nonsense as we rode. . . We hit the divide at the very top of the mountain in a tremendous thunderstorm. . . immense, huge, pounding rain. We sat under our horses for lunch and ate oranges, [and] were drenched. . . I was looking at Robert [. . .] and all of a sudden I noticed his hair was standing

---

[128] Smith and Weiner, p. 77; Goodchild, p. 16.

straight up [...] responding to the static." They arrived at Cowles at 7 p.m., fortunate not to have been struck by lightning on the mountain. After this idyllic romp in the Southwest, Oppenheimer headed across the Atlantic Ocean to devote himself to experimental physics.[129]

Not all was idyllic in England. Robert complained in a letter to Francis Fergusson: "I am having a pretty bad time. The lab work is a terrible bore, and I am so bad at it that it is impossible to feel that I am learning anything... The lectures are vile." Robert discovered that he held little talent for the physical preparations of slides or the setting up of scientific equipment. At one point, he had to coat glass slides with a thin coating for an experiment. Whether he could not get the coating to the appropriate thinness or that this activity quickly bored him, he was dissatisfied with this work. He did enjoy the fresh ideas and theories that spun around the labs and classrooms at Cambridge, but his mental state deteriorated throughout the fall semester.[130]

Over the Christmas break, Robert and Francis Fergusson (who was in England on a Rhodes Fellowship) traveled to Paris. They continued the practice they had begun at the Ethical Culture School of arguing over ideas and personal beliefs. This time though, Robert snapped. He jumped on Fergusson, grabbed him by the neck, and tried to strangle him. Francis quickly pushed Robert away, but this bizarre attack showed the stress that surfaced at Cambridge. In an apologetic letter to Francis several weeks after the blow-up, Robert says that he should come to Francis in "a hair shirt, with much fasting and show and prayer," to make up for the attack. Toward the end of the letter, Robert alluded to his "inability to solder two copper wires together, which is probably succeeding in getting me crazy."[131]

Others noticed that Oppenheimer was in trouble mentally. John Edsall, a classmate from Harvard who was also doing graduate work at Cambridge, observed: "There was a tremendous amount of inner turmoil, in spite of which [...] he kept doing a tremendous amount of work, thinking, reading, discussing things, but obviously with a sense of great inner anxiety and alarm." Robert confided with Edsall that he was seeing a psychiatrist.[132]

---

[129] Smith and Weiner, pp. 68, 80–81.
[130] Smith and Weiner, pp. 86–88.
[131] Smith and Weiner, pp. 91–92.
[132] Smith and Weiner, p. 92.

During the spring vacation, Oppenheimer joined Edsall and Jeffries Wyman in a tour of Corsica and Sardinia in the Mediterranean Sea. Although Robert at times complained of feeling depressed, the three friends had a good time exploring the island on foot, and Robert's mood improved. On the last night on Corsica before they were to go to Sardinia, Robert abruptly announced that he had to return to Cambridge. When pressed for an explanation, he said that he had left a poisoned apple on the desk of one of his professors back in England and had to return to make sure that the man was all right. Oppenheimer returned to England alone, to the great puzzlement of his two friends who chalked it up to Robert's inner turmoil. No professors died from poisoning that spring, and perhaps Robert merely needed to correct a mistake in some research that he had done for that professor.[133]

Oppenheimer in 1926

The holiday on Corsica did alleviate one of Oppenheimer's feelings of inadequacy. He met a woman on the island. Little is known of this mysterious relationship, but he wrote years later: "The psychiatrist was a prelude to what began for me in Corsica... What you need to know is that it was not a mere love affair, not a love affair at all, but love."[134]

Back in England, Oppenheimer confided to Edsall that he had been diagnosed with *dementia praecox*, now known as schizophrenia. Before Fergusson returned to the States that summer, he met Oppenheimer in front of the psychiatrist's office. Fergusson remembers the meeting as a turning point in Robert's difficulties: "I [saw him] standing on the corner, waiting for me, with his hat on one side of his head, looking absolutely weird. I joined him [...] and he walked at a terrific speed; when he walked his feet turned out [...] and he sort of leaned forward, traveled at a terrific clip. I asked him how it had been. He said [...] that the guy was too stupid to follow him and that he knew more about his troubles than the [doctor]

---

[133] Smith and Weiner, p. 93.
[134] Goodchild, p. 18.

did, which was probably true." A sign of Oppenheimer's intense intellect was that he cured himself of his emotional turmoil. "There's no doubt about it," Fergusson recalled, "Robert had this ability to bring himself up, to figure out what his trouble was, and to deal with it."[135]

Robert solved another difficulty that he grappled with that summer. He decided to abandon experimental physics, leave Cambridge, and move to Germany. The illustrious theoretical physicist Max Born had invited Oppenheimer to join him at the University of Göttingen. Just as Cambridge held a prestigious position in experimental physics, Göttingen was a world famous center for theoretical physics. The shift suited Robert's intellectual strengths. He recalled: "By the time I decided to go to Göttingen, I had very great misgivings about myself on all fronts, but I clearly was going to do theoretical physics if I could. . . I felt completely relieved of the responsibility to go back into a laboratory. I hadn't been good; I hadn't done anybody any good, and I hadn't had any fun whatever; and here was something I felt just driven to try." The year since he had left Harvard had taken its toll on Oppenheimer. He had changed from a chemist to an experimental physicist, and had abandoned that for theoretical physics. For a brilliant person trying to find his way, the time at Cambridge had not helped him, but he had high hopes for Göttingen.[136]

The fevered pitch of the work at Göttingen matched Oppenheimer's own inner fire and his work there helped advance quantum mechanics. Oppenheimer made his mark among the professors and graduate students at Göttingen, and others in the field began to notice him. With his Ph.D. in hand, Oppenheimer sailed to the United States from Liverpool in July 1927. He returned to New York, having made a name for himself at Göttingen. For this 23-year-old, he had weathered the crises in England, found his calling in Germany, and came back with knowledge and confidence. Physicists began to read the articles by Robert Oppenheimer with interest, and schools courted him to join their physics department. For someone who had his pick of prestigious universities to teach at, Robert's final choice showed the importance of the West to his life. Robert chose to join the physics departments in a joint appointment to the University of California at Berkeley and the California Institute of Technology at Pasadena.

---

[135] Smith and Weiner, p. 94.
[136] Goodchild, p. 18.

The impact of the West on Robert Oppenheimer is evident in both his choice of where to teach and with his selection of Los Alamos during the war as the site for an atomic weapons laboratory. He once said that two of the loves of his life were physics and New Mexico. From his first visit in 1922, New Mexico in particular, and the West in general, held a special influence on Oppenheimer's life. The impact of the West is seen in his letters as well as his actions. It was from the West that Oppenheimer launched the Atomic Age which transformed him and the 20th century.

GENERAL GROVES' INDISPENSABLE SCIENTIST

Robert S. Norris
Natural Resources Defense Council

General Groves tells us in his memoir, "Throughout the life of the project, vital decisions were reached only after the most careful consideration and discussion with the men I thought were able to offer the soundest advice. Generally, for this operation, they were Oppenheimer, von Neumann, Penney, Parsons and Ramsey."[137] I think it is correct to assume that these five are in ranked order with Oppenheimer at the top. Of course, Groves did not lack for scientific or technical advice. Groves' formal scientific advisers were James B. Conant and Richard C. Tolman. There were eight scientists who had already won Nobel Prizes working on the Manhattan Project and more than a dozen others would win the Prize after the war, one of them being Norman Ramsey.

Much has been written about the Oppenheimer–Groves relationship.[138] At the center of the relationship is Groves' decision, in the face of almost unanimous advice to the contrary, to choose Oppenheimer to head the scientific laboratory.[139] Groves was cautioned that Oppenheimer had no previous administrative experience. He had never been a department head or a

---

[137] Leslie R. Groves, *Now It Can be Told, the Story of the Manhattan Project* (Harper and Brothers, NY, 1962), p. 343.

[138] Robert S. Norris, *Racing for the Bomb: General Leslie R. Groves, the Manhattan Project's Indispensable Man* (Steerforth Press, South Royalton, VT, 2002), pp. 239–243 and passim; Gregg Herken, *Brotherhood of the Bomb: The Tangled Lives and Loyalties of Robert Oppenheimer, Ernest Lawrence and Edward Teller* (Henry Holt & Company, NY, 2002).

[139] "No one with whom I talked showed any great enthusiasm about Oppenheimer as a possible director of the project." Groves, *Now It Can be Told, the Story of the Manhattan Project*, p. 61.

Groves and Oppenheimer at
Ground Zero

dean. Luis Alvarez dismissed him by saying that he couldn't even run a hamburger stand. Most thought that an experimental physicist would be needed and Oppenheimer was a theoretical physicist. Unlike the other leaders of the fledgling project — Ernest Lawrence at Berkeley, Arthur Compton at Chicago, and Harold Urey at Columbia — Oppenheimer had not won a Nobel Prize, a factor that might cause less than full respect from his fellow scientists. To many, his political involvement with numerous leftist or communist causes was an obvious disqualification.

Groves was not convinced by any of these arguments. He apparently saw things in Oppenheimer that others at the time did not, qualities that would be essential to lead the scientific effort. Groves was a superb judge of character; it was one of the secrets of his success. He could size someone up quickly and determine whether the person was competent and capable to do the job that he wanted done. Groves had a "fatal weakness for good men," was Oppenheimer's immodest reply as to why the general had chosen him.

By the fall of 1942, Oppenheimer was already deeply involved in exploring the possibility of an atomic bomb. Throughout the previous year he had been doing research on fast neutrons, calculating how much material might be needed for a bomb and how efficient it might be. Under Compton's direction, Gregory Breit headed the theoretical group exploring these questions and Oppenheimer led the effort at Berkeley. After Breit resigned on 18 May 1942, Compton chose Oppenheimer as his replacement. He wasted no time and convened a summer study conference at Berkeley in Le Conte Hall, in July, to assess where the research stood. Oppenheimer referred to the group as "our galaxy of luminaries." The purpose of the conference, according to Robert Serber, was "to discuss the whole state of the theory, to make an independent assessment of whether the bomb was a reasonable possibility, and to discuss how well everything was known."[140]

---

[140] Robert Serber, *The Los Alamos Primer* (University of California Press, Berkeley, CA, 1992), p. xxx. The nine who attended were: Oppenheimer, Serber, Hans Bethe, Emil Konopinski, John van Vleck, Felix Bloch, Stanley Frankel, Eldred Nelson, and Edward Teller.

While there were uncertainties about such things as the cross sections and the number of neutrons per fission, they felt they were not enough to make the difference between success or failure.[141] The gun-assembly bomb design was a priority topic. At the time it was assumed that the gun design could use highly enriched uranium and plutonium. A second method of assembling the fissile material was also discussed. According to Serber,

> [Richard] Tolman came to me one day and talked about implosion... We discussed it that summer and wrote a memorandum on the subject... So the story of Seth Neddermeyer the lone genius coming up with implosion on his own is all hokum... It was Richard Tolman who brought the idea into the project.[142]

The group raised the alarming prospect of whether an atomic explosion might ignite the atmosphere. Oppenheimer was so concerned about this potential apocalypse that he traveled all the way from Berkeley to northern Michigan to see Arthur Compton at his summer cottage. As they walked along the beach Oppenheimer informed Compton of the potential dangers. Compton responded that the scientists would have to explore the matter further to arrive at a more definitive answer. They would have to reach a "firm and reliable conclusion" that it would not happen, otherwise the project would have to cease and the bombs could not be made. As he said, "Better to accept the slavery of the Nazis than to run a chance of drawing the final curtain on mankind!"[143]

In retrospect the range of topics covered at the Berkeley conference and the confidence the participants had about their conclusions is notable. It would provide a firm basis upon which to start at Los Alamos nine months later.

Oppenheimer came away from the conference with a better sense of what the next steps must be.

> [A] major change was called for in the work on the bomb itself. We needed a central laboratory devoted wholly to this purpose, where people could talk freely with each other, where theoretical ideas and experimental findings

---

[141] Serber, *Los Alamos Primer*, p. xxx.

[142] Serber, *Los Alamos Primer*, p. xxxii.

[143] Arthur Holly Compton, *Atomic Quest: A Personal Narrative* (Oxford University Press, NY, 1956), p. 128. The issue lingered over the next three years. James Conant's instantaneous reaction at the Trinity explosion was that something had gone wrong and that the atmosphere had ignited. James Hershberg, *James B. Conant: Harvard to Hiroshima and the Making of the Nuclear Age* (Alfred A. Knopf, NY, 1993), p. 760.

could affect each other, where the waste and frustration and error of the many compartmentalized studies could be eliminated, where we could begin to come to grips with chemical, metallurgical, engineering, and ordnance problems that had so far received no consideration. We therefore sought to establish this laboratory for a direct attack on all the problems inherent in the most rapid possible development and production of atomic bombs.[144]

General Leslie R. Groves

Oppenheimer was about to get his wish. Groves and Oppenheimer first met at a luncheon at President Robert Sproul's home, most likely on October 8. Groves had been on the job a little over three weeks and was visiting each of the universities where much of the research on the bomb was taking place. He had already made some big decisions and was ready to make more if the right men and ideas presented themselves. The two began to talk and they found that they had similar ideas on how to proceed. Oppenheimer was well versed on where the research stood; he knew the strengths of his fellow scientists and, as the above quote demonstrates, was primed for an all out effort. To the General all of these must have sounded like just what was needed. Perhaps he had found his man.

Groves returned to Washington. A second meeting was arranged for the following week, probably on the 15th or 16th. Groves had Oppenheimer meet him in Chicago. With many issues needing further discussion, Groves asked Oppenheimer to join him on the 20th Century Limited headed to New York. With Col. Kenneth D. Nichols and Col. James C. Marshall, the four squeezed into a tiny roomette and planned the laboratory as the train rolled eastward. On October 19th, Oppenheimer came to Groves' Washington office for further discussions. Sometime over the next week or two, Groves made up his mind about Oppenheimer.

What were those qualities that Groves discerned in Oppenheimer, insights that he was able to perceive earlier than the others? Of course in time

---

[144]U.S. Atomic Energy Commission, *In the Matter of J. Robert Oppenheimer* (MIT Press, Cambridge, MA, 1971), p. 12.

those qualities would reveal themselves and be recognized by almost everyone. His colleague Victor Weisskopf spoke of Oppenheimer as

> an unusually inspiring leader [who] had an extraordinary talent for grasping the essential points of a problem, even in fields far removed from his special training. His ability to be ready with the answer before one had finished formulating the question helped him to be aware of everything interesting that happened on the hill.[145]

> [At Los Alamos], He did not direct from the head office. He was intellectually and even physically present at each decisive step. He was present in the laboratory or in the seminar rooms when a new effect was measured, when a new idea was conceived. It was not that he contributed so many ideas or suggestions; he did so sometimes, but his main influence came from something else. It was his continuous and intense presence, which produced a sense of direct participation in all of us; it created a unique atmosphere of enthusiasm and challenge that pervaded the place throughout its time.[146]

The historians who wrote an official history of the project say that Oppenheimer "understood scientists, their methods, their prejudices, their temperaments. His professional stature, open manner, precision of thought, and articulate yet temperate speech equipped him admirably for the task ahead."[147]

Oppenheimer was sensitive to the needs of others, able to anticipate what it was they wanted. His charm and charisma had a powerful impact on almost everyone who met him. In addition to his obvious brilliance, Groves probably saw in the 38-year-old Oppenheimer a driving ambition. Groves often chose relatively young men for certain jobs. He felt that younger men were still hungry, full of energy and anxious to make a name for themselves as they pursued their careers.

Oppenheimer's formal responsibilities were detailed in a February 25, 1943 letter from Groves and Conant.[148] The desired goal was clearly

---

[145]Victor Weisskopf, *The Joy of Insight: Passions of a Physicist* (Basic Books, NY, 1991), pp. 132–133.

[146]Victor F. Weisskopf, "The Los Alamos Years," *Physics Today*, October 1967, p. 40.

[147]Richard G. Hewlett and Oscar E. Anderson, Jr., *The New World, 1939/1946: A History of the United States Atomic Energy Commission* (The Pennsylvania State University Press, University Park, PA, 1962), p. 230.

[148]David Hawkins, "Towards Trinity," in *Project Y: The Los Alamos Story*, History of Modern Physics, 1800–1950, Vol. 2 (Tomash Publishers and AIP, LA, 1983), pp. 495–497.

stated: "The laboratory will be concerned with the development and final manufacture of an instrument of war." The scientific director will be responsible for achieving that goal "at the earliest possible date" while maintaining secrecy "by the civilian personnel under his control as well as their families." He will rely on the advice of his scientific staff and keep Dr. Conant and General Groves informed.

Groves organized the Manhattan Project largely using the Corps of Engineers' model, which decentralized responsibility to area offices in the field, but kept direct links to them from headquarters. The headquarters of the Manhattan Project was Groves' office on the fifth floor of the New War Building at 21st and Virginia in the Foggy Bottom section of Washington, DC. From there he oversaw his vast empire with a surprisingly small staff. He said he took as his model Gen. William T. Sherman, who during his march to the sea limited his headquarters to what would fit into an escort wagon.

The Manhattan Engineer District offices were located in Oak Ridge, Tennessee, after initially being located in New York City, where it acquired its name. The District Engineer was Col. Kenneth Nichols, whose involvement in the project predated Groves.[149] The two largest units that Nichols administered were the Hanford Engineer Works and the Clinton Engineer Works (Oak Ridge). In terms of cost and number of personnel, Los Alamos was a small operation by comparison. The costs at Oak Ridge were about $1.2 billion, at Hanford almost $400 million, and at Los Alamos $74 million.[150]

From the outset Groves established dual lines of authority to Los Alamos: one to the military commander and the second to the scientific director, Oppenheimer. Both reported directly to Groves and did not go through the district engineer's office. With regard to Los Alamos, Groves assumed many of the functions of the district engineer and the area engineer. The direct lines permitted Groves (phone, Teletype, or frequent visits) to exercise broad

---

[149] From 1929 to 1931 Nichols served under Groves in an engineer battalion sent to Nicaragua to survey a prospective interoceanic ship canal.

[150] In current dollars. Hewlett and Anderson, *The New World*, 723–724. In today's dollars it would be approximately $24 billion, $8 billion and $1.48 billion, respectively.

control over the bomb program and to intervene in the day-to-day operations of the laboratory as needed. As he said,

> With respect to Los Alamos, it was directly my responsibility in every way, everything that happened. The orders were issued direct. We tried to keep Nichols informed to such extent as was necessary. So from a practical standpoint, though not on paper, the chain of command was direct from me to Dr. Oppenheimer.[151]

And again,

> Due to the magnitude of the District I retained personal direction of the Los Alamos bomb laboratory and took personal charge of the development of the weapon from the point where fissionable materials were supplied through and including the military operations.[152]

By my count, after the initial trip to choose the site in November 1942, Groves made two-dozen trips to Los Alamos between March 1943 and July 1945. Most of them were by train, traveling on the famous Santa Fe Super Chief out of Chicago, but a few were by plane. The trips to Los Alamos were often part of visits to other outposts of his empire. Oppenheimer came to Washington a few times during this period and on several occasions met Groves halfway in Chicago at the Met Lab or the area offices of the Corps of Engineers. Tolman and Conant visited Los Alamos periodically and reported to Groves on developments and problems. There was a steady correspondence between Groves and Oppenheimer in the form of memos, letters, and reports as well as scores of phone calls.

In short, the two worked together extremely closely, each believing the other was essential to the realization of their common goal of developing

---

[151] USAEC, *In the Matter of JRO*, pp. 171–172. According to Colonel Tyler's notes of a visit by the general, "General Groves reminded Col. Tyler that we are not in the chain of command under Oak Ridge, that we should absolutely take no orders from Oak Ridge, but should only look to them for whatever assistance that they can give us. At any time, when they give us orders which are in conflict with our policy, or which will interfere with us in anyway, we are to let the General know about it. We should send almost nothing through the District." Notes Taken During Visit of LRG, 7–9 February 1945, Folder 96, Manhattan Project Data, 1939–1948, Box 6, Nichols Papers, Office of History, USACE.
[152] LRG to Colonel T. D. Stamps, 21 August 1946, Folder 201, 1946, G.O.B., Box 2, Entry 7530C, Papers of LRG, RG 200, NARA.

and manufacturing an instrument of war to be used at the earliest possible date.[153] Whereas Groves was struck by Oppenheimer's many talents, so too was Oppenheimer admiring of Groves' abilities. One could not help but be impressed by Groves' decisiveness or the multiple responsibilities that he assumed and carried out simultaneously, including fissile material production, the creation of a military unit to drop the bomb, foreign intelligence activities, and international diplomatic negotiations. To Oppenheimer, Groves was, among other things, an expediter without equal. Anything that he needed, any piece of equipment, any scientist or expert was his for the asking. Groves was the master bureaucrat who knew which levers to push in Washington to get things done.

In their respective capacities Oppenheimer and Groves had to make dozens of decisions everyday on matters large and small. For truly crucial questions they worked together to solve some of the big issues.

Crash Implosion

For the first year or so there was a consensus among the scientists that both highly enriched uranium and plutonium could be used in a gun-type assembly bomb design, and the laboratory was organized to reflect that theory. In March 1944 the first half-gram of plutonium from the X-10 reactor at Oak Ridge arrived at Los Alamos. Previously, microscopic samples from the Berkeley cyclotrons were used for analysis. Experiments on the X-10 reactor sample suggested that it contained an isotope (Plutonium-240) that would fission spontaneously. This meant that pre-detonation would occur, a fizzle would result and that the gun method could not be used. It also meant that

---

[153] Oppenheimer wrote to Groves on 28 October 1946, "It is time that I should write to you a few words that have long needed saying, and that, as a matter of fact, I have promised to write you when the time came. There is no need for me to add words of appreciation for what you did during the war. Few men were in a better position to appreciate this than I. But the United States knows that it is in your debt, and will forever remain so." Groves responded on November 7, thanking him for his thoughtfulness, "I know that you know that the burden of responsibility our work placed upon me during the war was in no small way eased by the knowledge that I had your assistance in a key position. The confidence that I was always able to place not only in your scientific knowledge but also in your sound judgment on other matters was a constant source of comfort to me." Folder 201, 1946, G.O.B., Box 2, Entry 7530C, Papers of LRG, RG 200, NARA.

just possibly the entire Hanford project might be for naught, a $400 million investment wasted.

On July 17, 1944 Groves flew to Chicago to meet with Oppenheimer who had come from Los Alamos. Conant and Compton were at the meeting as was Enrico Fermi. As often happened during the Manhattan Project, a novel solution emerged just when it was needed. The answer was implosion, an idea first proposed two years earlier at the Berkeley summer conference but whose time was not yet ripe. Its time was ripe now. Groves and Oppenheimer gave implosion the highest priority and quickly reorganized the laboratory to turn the theory into a real bomb.

Thermal Diffusion

Among Groves' first decisions were to simultaneously pursue three paths to produce fissile material for a bomb: two isotope separation methods to enrich uranium and one method to produce plutonium. Two other ways of enriching uranium, by thermal diffusion and centrifuges, had been considered but were set aside. In the spring of 1944, Associate Director William S. "Deke" Parsons brought back to Los Alamos news of the success that the Navy was having with thermal diffusion. It dawned on Oppenheimer and others that if the different methods were integrated rather than operating independently then higher levels of enrichment might be achieved more quickly. The idea was presented to Groves. He agreed and quickly built the S-50 thermal diffusion plant at Oak Ridge. By the end of 1944, the slightly enriched uranium from S-50 was sent to K-25 and from there it went to Y-12 where levels above 80 percent U235 were achieved for the Little Boy bomb.

Bomb Production and Little Boy's Uranium

The Trinity test confirmed that the implosion design was sound. By that time the three Hanford reactors were producing plutonium in predictable amounts. Each Fat Man type bomb required six kilograms of plutonium. Groves had ensured the early delivery of enough plutonium through what was known as the "speed up" program. According to Dupont's schedule of October 1944, there would not have been enough plutonium available for the test until December 1945. This was totally unacceptable to Groves and he pressured Dupont to speed up plutonium production through various means, including increasing the power levels, pushing the uranium billets from the reactor more frequently,

"The Gadget" at Trinity Site

and shortening the time the billets remained in the cooling ponds. By these methods the schedule was advanced by six months and Groves' deadlines were met. The first six kilograms were available for the Trinity test and next six for the initial combat bomb that was dropped on Nagasaki.

The Hanford schedule permitted Groves to predict with some confidence how many bombs would be available over the next five months. How many bombs was it going to take to force the Japanese to surrender? Would these new super-weapons be used in an American land invasion in November, as George Marshall had wondered?[154] No one knew for sure and Groves saw as his job to keep producing bombs at a steady rate until the war was over. In his memoir Groves said that he thought that two bombs were probably enough to force the Japanese to surrender, though just after Trinity he felt that four might be necessary.[155]

There was serious consideration by some at Los Alamos, even including Oppenheimer, of using most or all of the sixty kilograms of the HEU from Little Boy and combine it with plutonium to make composite cores of the two materials. In this way more bombs could be built.[156] On July 19, Oppenheimer

---

[154] Barton J. Bernstein, "Eclipsed by Hiroshima and Nagasaki: Early Thinking about Tactical Nuclear Weapons," *International Security* 15(4) (Spring 1991) 149–173.

[155] "It is necessary to drop the first Little Boy and the first Fat Man and probably a second one in accordance with our original plans. It may be that as many as three of the latter in their best present form may have to be dropped to conform to planned strategic operations." LRG to Oppenheimer, 19 July 1945, Top Secret, File F.B.I.

[156] In a 1949 letter to Nichols, Groves comments on a misstatement that Robert Bacher had made recently in testimony to Congress: "That was his reference to the fact that the improved bomb used at Eniwetok had not been more than a gleam in someone's eye. He apparently has forgotten entirely Oppenheimer's urgent recommendation to me soon after July 16th, 1945 that we go directly to this model, instead of the previously planned unit scheduled for the second delivery." LRG to Nichols, 13 July 1949, Folder N, Box 6, Entry 7530B, Papers of LRG, RG 200, NARA. Groves was referring to Operation Sandstone in 1948, when composite core designs were first tested.

cabled Groves inquiring about which path to pursue in bomb production over the next three and a half months.

Groves told Oppenheimer that they should meet in Chicago to discuss the possible alternatives and schedules. While the details of the meeting remain classified it is clear that Groves made the decision not to use Little Boy's HEU to help make more implosion bombs with composite cores. Two bombs turned out to be enough to force Japan to surrender though a third one was ready to leave Los Alamos. It would have been ready for use on August 17 or 18 and a succession of others would have followed every ten days after that.

Ground Zero

The Manhattan Project is often held up as the paradigmatic case of how to mobilize talent and resources to achieve a goal effectively and quickly. In carrying out their responsibilities, the leadership manifested by Groves and Oppenheimer helped to establish the methods and procedures that have come to characterize successful large-scale collaborative efforts. One recent study examined those features.[157]

Start with smart talented people and have them produce something tangible as opposed to working on an abstraction or an idea. Young people are preferred as they are normally more energetic, confident, and curious and are more likely to work harder and longer. The undertaking has a better chance of success if it is driven by moral purpose. Put this special population in an isolated spot without any distractions. Living in Spartan conditions makes work the focus. The tendency to escape into the work may result in ignoring or not having the time to reflect on what is being produced. The cooperation of the many parts is essential toward realizing the overall goal. Ensure that those below have faith in their leaders, and make sure that the leaders have faith in those below. Though Americans like to believe in the triumphant individual that meets challenges and overcomes adversity, it is really a blend of individual

---

[157]Warren Bennis and Patricia Ward Biederman, *Organizing Genius: The Secrets of Creative Collaboration* (Addison-Wesley Publishing Co., Reading, MA, 1997).

and collective effort that gets things accomplished. The leader finds greatness in the group and also helps them find it in themselves.

Groves and Oppenheimer were two effective leaders who implemented these procedures and produced a bomb in a little over one thousand days. For that they were indispensable.

CHAPTER FOUR

Personal Reflections on Oppenheimer

OPPENHEIMER AS A TEACHER OF PHYSICS AND PH.D. ADVISOR

Edward Gerjuoy
Professor Emeritus of Physics, University of Pittsburgh

Introduction

I am delighted to have the opportunity to speak at this Symposium on the subject of Oppenheimer as a physics teacher and Ph.D. advisor. Before proceeding any further, however, I must offer an apology and a caution. The apology is to those members of my audience who are not physicists; I regret my inability to find a way, satisfying to me, to deliver this talk without referring to physicists and physics topics that probably are unfamiliar to non-physicists. The caution, to all members of my audience, is that there almost certainly are other still living students of Oppenheimer's who could speak more knowl-

Edward Gerjuoy

edgeably on my subject than I can. I say this in all sincerity and without undue modesty. For various reasons, some of which I shall address, Oppenheimer supervised the Ph.D. researches of most of his other students more closely than he supervised mine; also whereas several still extant Oppenheimer Ph.D.'s accompanied him to Los Alamos, I did not. But, very likely largely because I still am able to stand at a podium, here I am, and I will do my best to convey what learning and doing physics under Oppenheimer's tutelage was like.

Oppenheimer in the Classroom

I was enrolled as a graduate student in the Berkeley physics department from August 1938 to January 1942. Actually when I arrived in Berkeley I knew practically nothing about Oppenheimer beyond his name, even though I had come to Berkeley intending to get my Ph.D. with him. But I immediately became well acquainted with him via the introductory quantum mechanics course he gave that I took in my first semester at Berkeley. I took his electromagnetic theory course in my second semester, and in succeeding years took his advanced quantum mechanics and field theory courses. In each of these courses he manifested the same distinctive teaching style, many aspects of which merit detailed description.

- First, and most significantly, he obviously always came to class well prepared, although he equally obviously could have winged it with ease had he not devoted some advance time to planning what he intended to present. I would not say anything nearly as complimentary about the professors who gave any of the other non-Oppenheimer physics graduate courses I took at Berkeley. At least one of these other professors usually came to class unprepared and floundered at the board; the remainder were well prepared but, in contrast to Oppenheimer, did not always have the course subject matter at their fingertips and could be rattled by questions.

- Oppenheimer gave no final exams or any other tests. He did assign numerous homework problems, however, many of which were highly instructive and non-routine. These homework problems always were graded by Oppenheimer himself, again unlike the practice of a number of other Berkeley physics professors. I still have the homework solutions I submitted in his advanced quantum mechanics course, with his handwritten comments in the margin.

- He did not designate a textbook for any of his courses that I took, nor did he assign readings or homework problems from any textbook. In fact he only rarely explicitly cited any sources for the classroom material he presented. If we students desired alternative or otherwise clarifying presentations, we generally had to locate them on our own. I add that Oppenheimer's failure to assign a textbook in his electromagnetic theory course is revealing of his instructional bent. Much of the material he presented, though unquestionably classical electromagnetic theory,[158] unmistakably was intended to serve as an introduction to the newly formulated, indeed still-developing-at-the-time, quantum theory of radiation; such hypermodern material, though standard textbook fare today,[159] simply could not be found in any of the then-available electromagnetic theory textbooks. His failure to assign textbooks for his quantum mechanics courses is not revealing and requires no comment; at the time, barely a decade after Schrödinger's formulation of his wave equation, there weren't any English language texts for him to assign.[160]

- Each class hour literally was a lecture, delivered at high speed. The oral delivery was accompanied by numerous equations written on the board at correspondingly high speed, along with (when appropriate) equally rapidly performed, rarely erroneous calculations. The speed was such that the only way I possibly could grasp the material was to take hastily scribbled notes as he spoke, from which scribblings I would prepare more complete notes as soon as possible after the lecture, while it still was fresh in my mind; there was no textbook to consult, I remember. I am quite certain that every other serious student in those courses of Oppenheimer's that I attended did the same as I; indeed I remember numerous occasions when several of us would argue at a blackboard about precisely what he had imparted. Preparing these course notes took a lot of time; certainly I spent more time on each of Oppenheimer's courses than I did on any of the other courses I took in graduate school. On the other hand, I undoubtedly learned far more physics from each of Oppenheimer's courses than I did from any of the other graduate courses I took. I will add here that students who took his introductory quantum mechanics, and electromagnetic theory courses in 1940 or 1941 apparently were able to employ, as the equivalent of texts,

---

[158] See the notes to Oppenheimer's electromagnetic theory course, prepared c. 1939 by Shuichi Kusaka.

[159] See., e.g., J.D. Jackson, *Classical Electrodynamics* (Wiley, NY, 1962).

[160] For e.g., L.I. Schiff, *Quantum Mechanics* (McGraw Hill, NY, 1968), which is compatible with Oppenheimer's approach to quantum mechanics (doubtless because Schiff absorbed that approach during his 1938–40 tenure as Oppenheimer's research associate), was not published until after the end of World War II; H.S. Kramers, *Quantum Mechanics* (Dover, NY, 1964), which Oppenheimer lauded in his advanced quantum mechanics course, was not translated from German until 1964.

printed course notes prepared by several Ph.D. students of Oppenheimer's, but these course notes[161] were not available when I took his courses.

- I have no memory of him ever initiating any sort of Socratic dialogue with the class, nor do I recall him pausing in any calculation to ask the class for suggestions on what to do next. In so stating I am not implying that he would not take questions. If at any time during the lecture there was something a student didn't understand, said student could feel free to interrupt Oppenheimer with a question. I recall no indications that Oppenheimer minded such interruptions; rather he generally would answer patiently unless the question was manifestly stupid, in which event his response was likely to be quite caustic. Unfortunately his patient answers often were not illuminating; seemingly, Oppenheimer did not have the gift of putting himself in a student's place and recognizing that what was evident to him might not be evident to the student. A student who persisted after receiving Oppenheimer's initially patient answer could expect to find himself on the receiving end of the same sort of sarcasm that an obviously stupid question would elicit. I also must say, however, that I never saw any indications that he bore any grudges at students who momentarily had taxed his patience.

- I haven't yet mentioned probably the most distinctive feature of his lectures, namely his chain-smoking. He spoke quite rapidly, and puffed equally rapidly. When one cigarette burned down to a fragment he no longer could hold, he extinguished it and lit another almost in a single motion. I still can visualize him in his characteristic blackboard pose, one hand grasping a piece of chalk, the other hand dangling a cigarette, and his head wreathed in a cloud of smoke. I add parenthetically that according to information kindly furnished to me by the Berkeley physics department, of the 24 students who received their Ph.D.'s under Oppenheimer's direction during the years 1931 through 1943 (see Appendix A[162]), as many as ten still were

---

[161] B. Peters, "Notes on Quantum Mechanics. Physics 221, Oppenheimer 1939." Also see footnote 158 above.

[162] Appendix A lists the names and thesis topics of every Oppenheimer student who earned his/her Ph.D. at Berkeley. All but the first two of the names in Appendix A have been compiled by Raymond T. Birge, *History of the Physics Department* (University of California, Berkeley, CA), at Vol. 3, Appendix 13. Harvey Hall and John Franklin Carlson were omitted from Birge's compilation, apparently because he mistakenly had tabulated them as students of Professor William Williams (see Birge, *ibid.*, Vol. 4, Appendix 17), although Birge himself admits that "Williams never had a single research paper to his credit", *ibid.*, Vol. 2, ch. 7, p. 10. Actually Birge himself states that Carlson worked with Oppenheimer, *ibid.*, Vol. 14, ch. 11, p. 25.

alive as of April 2004, though I have every reason to think that they all are at least as old as I. I am including this fact in my talk so that Cindy Kelly, President of the Atomic Heritage Foundation sponsoring this Symposium, can recoup her expenses by conveying said fact, for an excessive fee of course, to the tobacco industry as conclusive evidence that the health risks of second-hand smoke have been vastly overrated.

- In my view these just-described aspects of Oppenheimer's classroom teaching style justify the following three conclusions, with which I shall close my discussion of his classroom teaching:

  ◆ First, although his primary interest as a physics professor surely was research not teaching, he nevertheless took his classroom teaching duties very seriously and performed conscientiously, even if sometimes impatiently.

  ◆ Second, he deserves credit for his painstaking efforts to construct unhackneyed courses that would lead students into productive physics research as rapidly as their native talents would allow.

  ◆ Third, there must have been many students who did not profit significantly from Oppenheimer's courses, whether because they lacked the necessary native talent and/or had poor undergraduate training, or merely because they were more interested in experimental physics than in theoretical physics and did not want to put in the large amounts of time required to adequately work up their class notes.

- This third conclusion is bolstered by the words of a quite talented physicist who has been a good friend of mine ever since we were fellow graduate students at Berkeley; wholly unlike me, however, he earned his Ph.D. in experimental physics and then worked at Los Alamos during the war. My friend has written (I quote):

  I did take E&M from Oppie, but not QM. I did not like Oppenheimer as a teacher... He was unaware, I think, of how little the students were understanding what he was talking about. As you know, he never gave any examinations and so he never really knew how much students had learned.

- I do not share my friend's reaction to Oppenheimer's classroom teaching, but feel there could be some truth in what he says; in any event I believe it is appropriate for you to hear not only my theoretical physicist conclusions, but also this assessment by a competent non-theorist.

Oppenheimer's Group

My classroom recollections of Oppenheimer, which I have just encapsulated for you, actually are considerably less vivid than the recollections stemming from my membership in his group of Ph.D. students. I joined Oppenheimer's group in roughly the spring of 1939. Joining his group was not a formal

procedure; as I recall I told him I was interested in obtaining my Ph.D. under his direction, and he merely said very well, we would see what I could do. He didn't immediately give me a research problem and I didn't expect him to; I had been at Berkeley less than two semesters and had not even taken his advanced quan-

Oppenheimer conducting a lecture

tum mechanics course. So at the time what joining his group required of me was nothing more than my seeing to it that I, like all the other students in his group, regularly showed up at the weekly theoretical physics seminar he ran. One of the first things I learned from attending Oppenheimer's seminar was that he did not mind being called Oppie; although Oppenheimer could be fearsome, he did not put on airs. I soon fell into the habit of thinking about him as Oppie, have done so ever since, and therefore will take the easy route of calling him Oppie in the remainder of this talk.

I will be saying a lot about Oppie's seminar, where so much of my shaping into a theoretical physicist took place. But first I should tell you about his group and say much more than I already have about Oppie himself. Oppie's group was composed: (i) of the students working for their Ph.D.'s under his direction; (ii) of various Ph.D. theorists, in residence at Berkeley for an extended period, who either explicitly had come to work with Oppie or who at least wanted to participate in his seminar. (I remember at least two theorists falling into this category, one of them a native Australian who had earned his Ph.D. in England.); and (iii) of his research associate, who already had his Ph.D. and today probably would be called a postdoc. Oppie had a research associate, who was supposed to help Oppie supervise his many students, every year I was at Berkeley; his research associate when I joined his group was Leonard Schiff.

The number of students in Oppie's group deserves special mention. Oppie joined the Berkeley faculty in the fall of 1929 and remained at Berkeley until the summer of 1942, when he was given a leave of absence to work on you-know-what. Of the 24 students who earned their Ph.D.'s under his supervision during this 13-year period (see Appendix A), fully 16 received their degrees after 1939 (see Appendix A); moreover I recollect that every one of these 16 already was a fellow graduate student by the fall of 1940, when I published my first paper. Recognizing that there also were some (though actually surprisingly few) students of Oppie's who never received their Ph.D.'s, the facts I have just recited imply that during every one of the years I was in Berkeley, Oppie was the Ph.D. thesis advisor of between 15 and 20 students, of whom a very large fraction eventually did earn their Ph.D. degrees.

These are remarkable statistics. Even in the most halcyon days this nation's physics departments have enjoyed since World War II, with would-be theorists flocking to U.S. graduate schools from the entire world, I doubt that there have been more than a handful of professors in this country who at any one time supervised as many ultimately successful theoretical physics Ph.D. students as Oppie did. Equally remarkable is how varied were the topics of those 24 Ph.D. theses he supervised between 1929 and 1942 (see Appendix A); they range from the interactions of electromagnetic radiation with matter (probably Oppie's favorite subject), to atomic spectroscopy, to nuclear structure, to neutron capture, to cosmic ray phenomena, to neutron stars and general relativistic gravitational collapse. In their totality this collection of Ph.D. topics, which spans just about all the subjects anyone would include under the heading "Modern Theoretical Physics" during the 1930s and early '40s, are a testament to the breadth and depth of Oppie's theoretical physics interests and abilities.

An obvious question now comes to mind: how did Oppie manage to secure so many students capable of earning a theoretical physics Ph.D.? The answer to this question is twofold, but not profound. First, his students were very good because before World War II popularized physics, the only students who even dreamed of attempting a Ph.D. in theoretical physics were those who had received very good grades in undergraduate physics; moreover those grades were earned in that archaic era when grade inflation had not yet been invented. The real question, therefore, is: why did so many of these very good would-be theoretical physics Ph.D.'s who received their undergraduate degrees in the decade or two before World War II come to Berkeley to work with Oppie? Indeed I would guess, admittedly without having tried to examine any relevant records, that the aforementioned 24 Oppenheimer Ph.D. recipients were a quite significant fraction of all the theoretical physics Ph.D. degrees earned in the U.S. during the years 1929 through 1943.

The answer to my second question stems from the fact that quantum mechanics, the ultimate foundation for most modern theoretical physics investigations, was developed in Europe and remained essentially arcane until 1926, when Schrödinger's formulation of his famous equation made quantum theoretical research accessible to non-geniuses like myself. Oppie was one of the very few American theoretical physicists who was both lucky enough to have learned quantum mechanics in Europe right around 1926, and talented enough to usably bring this learning back to the United States. He received his Ph.D. in 1927 from the University of Göttingen, having studied with Max Born, a very, very famous theoretical physicist; he joined the Berkeley physics department only two years later. In the years between 1929 and 1935, say, before so many great European physicists fled Hitler and began to establish their own modern theoretical physics research groups in this country, students who wanted to do research at the forefront of theoretical physics without going abroad enrolled in the Berkeley physics department to work with Oppie because there really were very few other professors in the United States actively engaged in such research. In 1937, therefore, when I was asking my undergraduate professors where I should go to do research in modern theoretical physics, the only established group they could point me to was Oppie's. That he was located in Berkeley, about as far from my family in New York as I possibly could get without leaving the United States made my forced decision to work with him a very happy one.

This brings me to Oppie himself, whom I have yet to properly describe. He was tall and absurdly thin. He also rarely was motionless; if nothing else he would be puffing on his cigarette or waving it around as he talked. He was well educated and well read; we all have heard of his ability to quote from the original Sanskrit. His face was mobile; how he was reacting was no secret. When I knew him he was between 35 and 40, and doubtless still at the peak of his physical and mental powers. Those mental powers surely were prodigious. He had a lightning quick mind, was tremendously verbal, and pretty obviously always found the words to say exactly what he wanted to say; unfortunately when he was talking physics these words did not always convey to his audience what he hoped to convey, as I already have indicated.

His relations with his students, including me, were surprisingly informal. For example, he allowed his students to drop into his office at any time to consult physics books in his personal library which were not available in the department library; I didn't even have to knock. His office was deep, moderately wide, and quite bare; except for the bookshelves and a blackboard running the length of the room I recall only a single desk and chair, which I hardly ever saw him using. He did not have any regular office hours. He could be moody and often seemed to brood; if I found him alone in his office, or elsewhere for that matter, his demeanor instantly made it apparent whether or not I should dare speak to him. But if he was willing to talk, or already was speaking with another student, I did not hesitate to ask him a question right then about some physics matter bothering me; there was no need for an appointment. Assuming one did catch him willing to be disturbed, this just-described easy accessibility didn't mean that I or any other graduate student thereupon, freely would toss physics questions at him without forethought. As I also already have indicated, his reaction to a question he deemed stupid tended to be very caustic; one was likely to depart his company quite depressed.

Concerning Oppie's knowledge and understanding of physics, Raymond T. Birge, physics department chair during my years in Berkeley, has written:

> I can testify, without I think fear of contradiction, that no matter what field of physics was being discussed, Oppenheimer had more numerical facts in his head and ready for instant application than did any experimental physicist present, even including those working in the particular field being discussed. Such remarkable knowledge of experimental facts, plus the ability to marshal them effectively into theory, is in my opinion what constitutes genius.[163]

My own evaluation would not be quite as overwhelmed. While Oppie's knowledge of both theoretical and experimental physics undoubtedly was extraordinary, I would judge that more than a handful of physicists in his generation matched that knowledge; Enrico Fermi and Hans Bethe, both of whom won Nobel Prizes, immediately come to mind. I also judge most physicists would agree Fermi was a genius, but would not confer that appellation on Bethe, despite Bethe's remarkably prolific career; similarly I would not term

---

[163] Birge, *ibid.*, Vol. 3, ch. 9, p. 31.

Oppie a genius. Although he, together with his students, published a very considerable number of important theoretical physics papers, his comprehensive understanding of known physics, especially the startling ease with which he rapidly grasped the newest developments, seemingly were not paralleled by any completed research that deservedly should be termed unusually creative, for example, worthy of the Nobel Prize.

It is possible that this just stated assessment of Oppie's talents as a physicist, which I believe is in keeping with conventional opinion, overcautiously underestimates his creativity. Less than a month ago a publication of the American Physical Society[164] pointed out that in 1939 Oppie and his student Hartland Snyder, a member with me of Oppie's group, published a long unnoticed paper showing that black holes could exist. This paper well might have earned Oppie a Nobel Prize, had he lived long enough to see the existence of black holes confirmed by astrophysical observations. In any event I want to emphasize and re-emphasize that nothing I have said should be taken to imply I have anything other than the greatest admiration of Oppie's talents as a physicist, in all aspects of those talents. In particular, speaking as a former student of his, I fully endorse the statement made in a recent biographical sketch, that as a professor at Berkeley Oppie "became arguably the most important and certainly the most charismatic physics theorist in the United States."[165]

Birge also writes: "Like all geniuses, Oppenheimer was very absent minded."[166] As evidence for this asserted absentmindedness, Birge recounts a story involving Oppie and Melba Phillips, the third student and the only woman to earn a Ph.D. under Oppie's direction (see Appendix A), with whom Oppie wrote a quite famous paper on the physics of deuteron collisions with nuclei. I must say that I never saw the slightest evidence of any absentmindedness in Oppie; quite the contrary in fact. Furthermore it is inconceivable

---

[164] The *Physical Review Focus*, a monthly email publication of the American Physical Society, wrote on June 1, 2004: "Had J. Robert Oppenheimer not led the U.S. effort to build the atomic bomb, he might still have been remembered for conceiving of black holes. His 1939 Physical Review paper (*Phys. Rev.* 56, 455), written with graduate student Hartland Snyder, described how a star might collapse into an object so dense that not even light could escape its gravitational clutches. The paper was hardly noticed until the 1960s, when astrophysicists began to seriously consider that such extreme objects might exist."

[165] "Oppenheimer", biographical article by B. Bederson, *Encyclopedia of Science, Technology, and Ethics* (Macmillan, 2005).

[166] Birge, *ibid.*, Vol. 3, ch. 9, p. 31.

to me that any absent-minded theoretical physicist could have successfully dealt with the myriad of details that unavoidably required Oppie's handling during his justly famed and universally admired wartime directorship of the Los Alamos nuclear weapons project.

Nevertheless I will repeat here this story about Oppie and Phillips, which dates from 1934, because it does shed a rather different light on Oppie than is suggested by anything I have said so far, and because it still was being snickered about when I joined Oppie's group in 1939. Briefly, at 4:00 a.m. one morning, a policeman patrolling the Berkeley hills found Melba in a panic, alone in Oppie's parked car. She said she and Oppie had been sitting there for some time when, about two hours previously, Oppie had excused himself. Oppie had not returned. The police looked for him everywhere in the vicinity of the car, and then telephoned the Faculty Club where he was lodging at the time. The Faculty Club found him in bed, asleep.

According to the newspaper stories about this incident (a melange of which had been passed on to me by some graduate student whose name I forget, and which I still possess in yellowed form), Oppie told the police that after leaving the car he simply had forgotten about Melba and had gone home. The incident speaks for itself and requires no comment from me. I will say, however, that because Melba got her Ph.D. in 1933 (see

San Francisco Chronicle, 14 February 1934

Appendix A), well before this abortive tryst occurred, Oppie cannot be accused of having violated any of the modern precepts against intimacies between a male professor and a female student working under his direction.

Oppie's Seminar

Returning now to Oppie's seminar, its regular attendees primarily were the members of Oppie's group, as well as Oppie himself of course. But there were other more or less regular attendees during my years at Berkeley. Of such attendees the most noteworthy was Stanford Professor Felix Bloch, a man of just about Oppie's age, who approximately once a month would drive in from Palo Alto with a few of his students. That drive, which can be wearying even now, was far more wearying then, before the freeways and — in my first months at the seminar — even before the Bay Bridge; thus Bloch's attendance also was a testament to Oppie's theoretical physics interests and abilities. Another noteworthy attendee, primarily because I never was able to find out why she attended, was the mysterious Madame Kokshoreva, who to my boyish eyes seemed to be 50 years old. She was not a graduate student nor had any other connection with the university to my knowledge; nor did I ever see the slightest sign she had any inkling about any subject the seminar was discussing. In fact I can't remember her asking a question or even uttering a word other than answering to her name. But she attended the seminar quite faithfully, and Oppie never seemed to bat an eye when she entered or when she left.

The seminar was Oppie's domain, his fiefdom. He selected the speakers; except on rare occasions he totally dominated its proceedings. In complete contrast to his classroom practices, Oppie almost never allowed himself to be the seminar speaker. He preferred instead to sit in the front row and interrupt the speaker with questions, except that not infrequently he became so exasperated by the answers he was receiving that he shot to the front of the room and took over the blackboard, thereupon morphing into his classroom lecturing mode. Unless he formally had scheduled a speaker from outside his group, Oppie's first choice for speaker always was whomever prominent theoretical physicist momentarily happened to be visiting the Berkeley physics department; a scheduled talk by any member of Oppie's group obviously could be and would be postponed.

In those years the Berkeley cyclotron, which in 1939 earned its professor inventor Ernest Lawrence the Nobel Prize, was one of the seven physics wonders of the world. Famous physicists, both experimentalists and theorists, flocked to Berkeley from all corners of the globe; for example, to name just two such visitors who had made important theoretical physics contributions: 1938 Nobelist Enrico Fermi gave an extended series of lectures in 1940, and 1945 Nobelist Wolfgang Pauli visited in 1941. Oppie more often than not managed to convince such visiting theorist-targets-of-opportunity to speak in his seminar. The results were exciting and instructive to all of us student attendees; we were able to hear about research at the forefront of theoretical physics right from the horse's mouth, and (perhaps even more importantly) could see for ourselves that not all great theoretical physicists were cut from Oppie's mold.

Of course Oppie wasn't able to secure visiting target-of-opportunity speakers every week; prearranged seminar speakers from outside Berkeley came not infrequently, but also not frequently. All in all I would judge that well over half the time the seminars were delivered by members of his group. Students would speak on research they had completed and were about to write up; occasionally Oppie would assign someone to talk on a published paper he thought worth discussing. When student speakers in such categories could not be mustered, the duty of speaking would fall back on Oppie's research associate. It was Oppie's practice each year to assign his research associate a broad subject, almost the equivalent of a course, which said associate would speak on in a continuing fashion, whenever no other speakers were available; in this fashion, Oppie and the other seminar attendees, whether students like myself or Bloch from Stanford, could be sure that a speaker of some sort, whether visitor, student or research associate, always would show up.

I already have said that tossing questions at the speaker was Oppie's preferred seminar role. He fell into this role with visiting and home-grown speakers alike. Especially with home-grown speakers, though by no means necessarily, if a question was not answered to Oppie's satisfaction he would furnish his own answer; moreover as I said he was not averse to brushing the speaker aside and going up to the blackboard if he felt the occasion warranted the intrusion. In this question-answering mode he did not distinguish between his own questions and the questions of others, nor did he treat Bloch's questions any differently than those from any other seminar attendee. Unfortunately, much as in his classroom performances, his answers often did not wholly clarify the issues at hand, even though his seminar attendees obviously were much more sophisticated theorists than the students in his courses. I well remember the many occasions when, after one of Oppie's answers, the cry, "But Oppenheimer! . . . ," uttered in an unmistakably German accent, welled up from Bloch. We students in Oppie's group, though no less baffled than Bloch by what Oppie had said, reveled in Bloch's discomfiture and were fond of saying that Bloch was Oppie's most advanced student. It was not until after the war, when I had begun to teach physics myself, that I realized Bloch, though slow thinking in comparison to Oppie's superfast mind, was a distinguished physicist who otherwise hardly would have had a professorship at Stanford. Indeed Bloch won a Nobel Prize in 1952.

As I recall, Oppie's seminar performances avoided disconcerting any of his visiting speakers; Oppie basically was a polite man, and I judge he took care not to be impolite to these visitors, who after all were doing him a favor. Unfortunately I can't say the same about his questioning of his students when they spoke. With them his questioning was fierce, often cruelly so. I went through a couple of speaking assignments without being too badly scathed; many of my fellow students were not as lucky. I want to emphasize here that I do not believe Oppie was in any way sadistic; on the contrary my overall assessment of his behavior leads me to think he legitimately could be termed kindhearted. Furthermore I feel confident that the questions Oppie put to his student speakers were designed not to embarrass but to elucidate, more often for the benefit of the audience than for himself. In fact I wouldn't be surprised if Oppie's persistent questioning was nothing more than an automatic attempt to remedy the discomfort he clearly felt when hearing any theoretical physics statements he thought wrong or even imprecise; it could have been like scratching an itch. Sadly, however, Oppie appears to have somehow lacked the empathy that would cause him to draw back, once his previous questions had reshaped the student speaker into a quivering hulk incapable of profiting from, much less answering, any further questions.

Nor, when Oppie's research associate Leonard Schiff gave a seminar, did Oppie treat Schiff any more kindly than he treated his student speakers. Schiff had been assigned the task of taking the group through a famed treatise on the quantum theory of collisions between charged particles, published some years earlier by two British theoretical physicists named Mott and Massey.[167] The contents of this treatise, especially the mathematics it employed, are standard fare today, but they were far from standard when Schiff was working his way through the book. So it was quite understandable that Schiff might not fully understand the mathematical properties and physical implications of all the equations from Mott and Massey he would reproduce on the board during any one seminar. Nevertheless Oppie, who perfectly legitimately systematically asked Schiff searching questions about each and every equation Schiff wrote down, couldn't seem to halt his questioning and let Schiff go on to another

---

[167] N.F. Mott and H.S.W. Massey, *The Theory of Atomic Collisions* (Oxford, 1933).

equation when Schiff's answers clearly showed that he needed more time to work his way through the equation Oppie was asking about. On more than a few occasions Oppie had Schiff, who was a gentle soul, visibly on the verge of tears. Much as was the case with Bloch, Oppie's treatment of Schiff left his students with no real appreciation of Schiff's talents. Certainly we would not have predicted that Schiff would have a distinguished career, including chairing the Stanford physics department.

I cannot refrain from contrasting Oppie's treatment of Schiff during Schiff's seminars with Oppie's corresponding treatment of Julian Schwinger, who in 1940 replaced Schiff as Oppie's research associate. All Oppie's students, including myself, were eagerly anticipating Julian's first seminar. But whereas the other students were wondering how long it would take Julian to shrivel under Oppie's questioning, I was wondering how Oppie would react to Julian's refusal to shrivel. These differing wonderments stemmed from the fact that I, unlike every one of Oppie's other students, had been exposed to Julian's talents during my undergraduate years at City College in New York. I do consider Julian, who shared the 1965 Nobel Prize for developing the very important modern formulation of quantum electrodynamics, to have been a genius. In fact my greatest claim to fame may be that when Julian and I were in the same mechanics class in City College I got an A whereas he only earned a B, I guess because he angered the instructor by getting 100% in all the exams even though he never attended the regular class sessions after the first week.

Anyway Julian's first seminar, on a subject I no longer remember, went exactly as I expected, but totally astonished every other student of Oppie's. Namely Julian started talking and very soon Oppie, in accordance with his usual practice, asked Julian a question, which Julian answered. Another question followed very shortly thereafter, and Julian answered. More questions came; more questions were answered. After about a dozen questions, answered by Julian with no visible sign of distress whatsoever, Oppie stopped firing questions and let him finish his seminar essentially without further interruption. Nor did he ever again unduly interrupt during any succeeding seminar of Julian's.

I stress that these contrasting treatments of Schiff and Schwinger should not be taken to imply that Oppie stopped asking Schwinger questions because he saw he couldn't bully Schwinger the way he had bullied Schiff. In the first place I do not believe Oppie was trying to bully Schiff; he just lacked the empathy to recognize how badly he was distressing Schiff, as I already have explained. Furthermore Schwinger, though perhaps not as gentle as Schiff, was not a specially tough soul; if Oppie had wanted to bully Schwinger, he easily could have found a way. Oppie stopped asking Schwinger questions because it became apparent, in that first seminar and in Oppie's subsequent conversations with Julian, that Julian always knew what he was talking about and would sufficiently discuss any subtleties inherent in his seminar subject without having to be prodded by Oppie.

Closing Remarks

The foregoing said, it's time to start winding down this talk. I would have liked to summarize for you what and how Oppie's Ph.D. students learned from him outside his courses, but I have had to abandon this endeavor as beyond my powers. A major part of my difficulty is that, as I intimated at the very outset, Oppie did not closely supervise my Ph.D. research; in fact it would be quite accurate to assert that he hardly supervised my research at all. I need not go into a lot of details on this point. Suffice it to say that my thesis consisted of three completed papers, the first two written by me alone and the third a joint paper with Schwinger. Each of the first two papers arose out of problems initiated by experimentalists, who approached Oppie asking whether their data were consistent with theory. In each of these cases Oppie, after assigning me the problem of determining whether or not the experimental results were predictable, told me that I first should bring any questions I had to Schiff; I should feel free to bother Oppie only if Schiff couldn't help.

As it happened I didn't need to bother Oppie, nor did I have to bother Schiff very much. Moreover neither problem was at the forefront of theoretical physics, as I freely admit; also the data turned out to be quite predictable, as should have been expected. Consequently Oppie showed little interest in these researches and never asked me to brief him on how I was getting along; nor did he do more than barely skim the calculations I performed and the prepublication drafts of the papers I wrote. Here Oppie was willing to rely on Schiff, with only a bare minimum of his own checking; Schiff had carefully looked at my calculations and drafts, of course. I never even discussed with Oppie any aspects of the research I carried out for my third paper, the one I wrote with Schwinger; by this time Oppie completely trusted Julian. I did have the benefit of working very closely with Schwinger on my third thesis problem; in fact we toiled together night after night into the wee hours, performing all the required calculations in each other's company, sometimes independently, sometimes jointly at the blackboard. But working closely with Julian need not have been, and probably wasn't, anything like working closely with Oppie.

With the bulk of his students, however, Oppie was fairly closely involved with their Ph.D. researches. He was interested in many of the problems they were working on, and in not a few instances himself worked on the problems alongside his students, much as Julian worked alongside me. Those students therefore had a much better opportunity than I to see how Oppie's mind worked and to savor his approach to physics. Consequently I feel I must confine myself to imparting the one all-important benefit I believe all Oppie's Ph.D. students gained from their asso-ciation with him whether or not they

J. Robert Oppenheimer

worked closely with him, a benefit they might not have derived from a differ-ent Ph.D. advisor, even a very famous one; but before embarking on even this much less ambitious venture I have to renew the caution with which I started this talk.

Obviously all of us who earned our Ph.D.'s with Oppie learned a great deal of modern physics, in the courses we took from him and while working on the research problems he assigned. Even if he wasn't terribly interested in the outcomes of some of those researches, the problems all were nontrivial and fully involved modern physics; any student who completed one of those assigned research problems was transformed thereby, into a significantly more competent theoretical physicist than when he had begun the work. Please recognize, however, that what I have just said is in no way different from what any Ph.D. student of any of our great modern theoretical physicists would be expected to say. For instance, sticking with names I already have mentioned in this talk, I would expect that in discussing their mentors my words would be echoed by Fermi's students, by Bethe's, by the students of Max Born including Oppie himself, and by Schwinger's students, who eventually comprised 73 Ph.D.'s including three future Nobel Laureates.[168]

Even if it is true that Oppie's own published contributions to physics did not quite match those of the physicists I have just named, terming Oppie a mentor equal to any of them is a high tribute. Although I cannot really know, of course, I have every reason to believe the tribute is well deserved. But I also have every reason to believe Oppie's mentoring of his Ph.D. students deserves an additional tribute, one that I think probably should not be as readily shared with his great contemporaries, although again I cannot really know. I feel Oppie did his physics, talked about his physics, lived his physics, with an unusual passion, which had to inspire his students; in any event he sure inspired me. To give you just one of many possible illustrations, it bothered him, it tore at him, that he didn't understand how the pi mesons which in nuclei were so strongly interacting penetrated the earth's atmosphere so readily. Maybe he should have hit upon the idea that the mesons reaching the earth's surface really weren't pi mesons, but instead were other weakly interacting mesons — those we now term mu mesons; but since he hadn't conceived of mu mesons he couldn't stop talking about the anomaly that atmospheric penetration by pi mesons represented, in seminar after seminar and in less formal conversations with groups of his students.

Thus, and I now am concluding what I have to say, despite Oppie's sometimes overly ferocious questioning, despite the sarcasms that Oppie really should have suppressed, we his students respected him and felt indebted to him; knowing that Oppie so obviously passionately loved doing physics, that he so obviously always had physics in the forefront of his mind, helped us believe that becoming a competent theoretical physicist was worth the fairly enormous effort required, especially in those prewar economically depressed days when the word physics had no popular resonance and jobs for theorists were very hard to come by. And, for imbuing me with this belief, I respect and feel indebted to him still.

---

[168] See Jagdish Mehta and Kimball A. Milton, *Climbing the Mountain. The Scientific Biography of Julian Schwinger* (Oxford University Press, 2000).

Ph.D. Candidates Supervised by Robert Oppenheimer with date of Final Examination and Title of Thesis

(Compiled from Raymond T. Birge's *History of the Physics Department*, University of California, Berkeley, CA.)

1. *Harvey Hall, August 17, 1931*
   The Relativistic Theory of the Photoelectric Effect

2. *John Franklin Carlson, April 30, 1932*
   The Energy Losses of Fast Particles

3. *Melba N. Phillips, May 6, 1933*
   Problems in the Spectra of the Alkalis:
   A. Photo-Ionization Probabilities in Atomic Potassium
   B. Theoretical Considerations on the Inversion of Doublets in Alkali-like Spectra

4. *Arnold T. Nordsieck, May 11, 1935*
   The Scattering of Radiation by an Electric Field

5. *Glen D. Camp, May 14, 1935*
   Is a Relativistic Non-Conservative Theory of Mechanics Possible?

6. *Willis E. Lamb, Jr., April 18, 1938*
   I. On the Capture of Slow Neutrons in Hydrogenous Substances
   II. On the Electromagnetic Properties of Nuclear Systems

7. *Samuel B. Batdorf, April 23, 1938*
   An Investigation into the Penetrating Possibilities of Charged Particles of Arbitrary Magnetic Moment

8. *Sidney M. Dancoff, September 13, 1939*
   Three Problems in Quantum Mechanics:
   A. Virtual State of He^5 and Meson Forces
   B. On Radiative Corrections for Electron Scattering
   C. The Calculation of Internal Conversion Coefficients

9. *George M. Volkoff, February 27, 1940*
   I. The Equilibrium of Massive Neutron Cores
   II. The Oppenheimer–Phillips Process

10. *Philip Morrison, April 23, 1940*
    Three Problems in Atomic Electrodynamics:
    A. Internal Conversion of Gamma Rays of Arbitrary Multipole Order
    B. Internal Scattering of Gamma Rays
    C. Energy Fluctuations in the Electromagnetic Field

11. *Hartland S. Snyder, April 25, 1940*
    Five Problems:
    A. Cascade Theory
    B. Quadratic Zeeman Effect
    C. Gravitational Collapse
    D. Mesotron Collisions
    E. Energy Levels of Fields

12. *Joseph M. Keller, September 13, 1940*
    Precise Determination of the Fine Structure Constant from X-Ray Spin Doublet Splitting

13. *Robert F. Christy, April 22, 1941*
    Cosmic-Ray Burst Production and the Spin of the Mesotron

14. *Eugene P. Cooper, August 25, 1941*
    Three Problems in Quantum Theory:
    A. Internal Scattering of Gamma-Rays
    B. The Ground States of Be10 and C10
    C. On the Separation of Nuclear Isomers

15. *Shuichi Kusaka, February 3, 1942*
    Studies on the Spin of Elementary Particles:
    A. Electric Quadrupole Moment of the Deuteron
    B. The Interaction of Gamma-Rays with Mesotrons
    C. Burst Production of Mesotrons
    D. Beta-Decay with Neutrino of Spin 3/2
    E. Quantization of the Wave Field for Particles with Higher Spin

16. *Richard R. Dempster, March 10, 1942*
    I. The Calculation of Transition Probabilities for Photoionization of Sodium from the 3P-State

II. The Influence of Fluorescence upon the Central Intensities of the Solar D-lines

17. *Roy Thomas, April 13, 1942*

   Three Problems in Quantum Mechanics:
   A. Internal Pair Production in Radium C'
   B. Burst Production by Mesotrons of Spin One-Half and Zero Magnetic Moment
   C. High Energy Proton–Deuteron Scattering

18. *Eldred C. Nelson, April 24, 1942*

   Mesotrons and Nuclei:
   A. Internal Conversion of Gamma-Radiation in the L. Shell
   B. Isomeric Silver and the Weizsäcker Theory
   C. Note on the Neher–Stever Experiment
   D. The Ground States of Be10 and C10
   E. Pseudoscalar Mesotron Theory of Beta-Decay

19. *Bernard Peters, May 8, 1942*

   A. Deuteron Disintegration by Electrons
   B. Scattering of Mesotrons of Spin 1/2

20. *Edward Gerjuoy, May, 1942*

   Problems in Atomic and Nuclear Spectroscopy:
   A. On the Angular Distribution in the Reaction F19(p, alpha)
   B. Interference in the Zeeman Effect of Forbidden Lines
   C. Tensor Spin-Orbit Forces in H3 and He4

21. *Stanley P. Frankel, July 20, 1942*

   Three Problems in Theoretical Physics (the first of the classified research project theses)

22. *Chaim Richman, April 13, 1943*

   A Problem in Theoretical Physics (classified)

23. *Joseph W. Weinberg, June 4, 1943*

   Studies in the Quantum Field Theory of Elementary Particles

24. *David J. Bohm, June 5, 1943*

   Theoretical Investigation of Scattering of Various Nuclear Particles (classified)

25. *Leslie L. Foldy, June 15, 1948*
    Four Studies in Theoretical Physics:
    A. The Theory of the Synchrotron
    B. Theory of the Synchro-Cyclotron
    C. On the Meson Theory of Nuclear Forces
    D. The Energy–Momentum Relation for Particles Interacting with Fields

26. *Harold W. Lewis, June 16, 1948*
    Three Problems in Theoretical Physics:
    A. The Multiple Production of Mesons
    B. On the Reactive Terms in Quantum Electrodynamics
    C. On the Analysis of Extensive Cosmic Ray Shower Data

27. *Siegfried A. Wouthuysen, June 16, 1948*
    Self-Energy and Covariance in Field Theory

REMEMBERING OPJE: TEACHER, MENTOR, SCIENTIST AND FRIEND

David Pines
Institute for Complex Adaptive Matter, University of California Office of the President, Oakland, California

I would like to thank Cindy for this opportunity to pay tribute to Robert Oppenheimer as a teacher, mentor, scientist, and friend. My talk will be a series of snapshots of Opje (the Dutch spelling of the nickname he was given in Leiden) and his wife, Kitty, as I knew them in Berkeley, Princeton, and Paris. Through these, I'll try to give you a sense of Opje and his extraordinary capacity as a teacher. He was a truly charismatic person and communicator, a caring and inspiring mentor, a fascinating and witty friend, and a great scientist. I

David Pines

hope these vignettes will complement the portrait that has emerged through the many fine talks we have heard today.

I was an undergraduate and one-semester graduate student at Berkeley during the war, from August, 1941 to June, 1944, before going off for two years in the Navy. When I became a physics major in 1943, through those of his students on the Berkeley faculty who were still around (Joe Weinberg, whose lectures on quantum theory led me to think about becoming a theoretical physicist, and Dave Bohm, who gave guest lectures on plasmas in Weinberg's class on electricity and magnetism), I became acquainted with the Oppenheimer legend. Although he was not present and no one would say exactly where he was, he was a dominant figure with whom you could only be eager to study theoretical physics in Berkeley. There were numerous stories about him, some of which Ed Gerjuoy has recounted here today. His students and postdocs did not simply worship him, they did their best to imitate him. Thus we were told that Julian Schwinger, who was comparatively pudgy and short, tried to look, walk, and talk like Robert Oppenheimer, who was tall and thin and elegant. Julian even went so far as to wear a pork pie hat.

Opje in person in 1946 did not disappoint. He was also not exactly comfortable at this point with the fame that came his way. That fall when I returned to graduate work at Berkeley, I was introduced to him on the stairs going into LeConte Hall, the physics building. Just then, several people who were walking by stopped and stared. At which point, he turned on them and said, "You look funny, too."

Opje's course on quantum mechanics was justly regarded as a classic. His lectures were inspiring and, as one listened to them, seemed very deep, yet also clear. Looking back, I cannot imagine a better way to become acquainted with its triumphs, intricacies, and paradoxes.. Opje's course was derived in no small part from Pauli's article on quantum mechanics in the Handbuch der Physik. His lectures were challenging, refined, and rich in texture. We now possess not one, but two written accounts of them in the form of excellent books written by his students Leonard Schiff and David Bohm, Although both books were based on those lectures, they could not be more different. Bohm captures the philosophy and the poetry; Schiff the formalism. And yet each caught a sense of what Oppenheimer tried to convey.

He always began the lectures with an account of the famous exchange between Bohr and Einstein in 1927 at the Solvay Congress. He made that seem so simple and lucid but also so strange that he conveyed the essence of quantum mechanics. If you like, all the rest were details.

He was by then a very popular lecturer, his course a must for all graduate students in physics. Part of it was his extraordinary personae — his piercing blue eyes and striking physical appearance, his way with language and cigarettes or pipe. He had an instinct for the perfect phrase. *Time magazine* did a cover story on him about that time; it included his definition of what theoretical physics is — "What we don't know, we explain to one another." At the time I thought that was a casual remark. Through the years, I have realized that it encapsulates what theoretical physics is all about.

His parties were memorable. Although I was only a beginning graduate student I was fortunate to have been among those present at a number of them during 1946–47, as Opje and Kitty kept up the tradition established by Opje before the war, of inviting a mixture of the very young and the more senior of his colleagues and friends.

One aspect of their parties was the unusual ratio of drink-to-food. It was so out of proportion that at an Oppenheimer cocktail party you could easily consume anywhere from four to ten martinis while having only a little bit to nibble on from time to time. The result was predictable; most of his guests were thoroughly soused by the time the party was well under way.

Kitty had a fondness for formal parties as well. So for winter solstice that year, she organized a formal party on December 21, to greet Opje on his return from a trip to Washington. That might have been the only time after the war that his friends Ernest Lawrence and Haakon Chevalier were in the same room, both looking quite elegant in their tuxes.

In the spring of 1947, his students in quantum mechanics realized he was going to give a lecture on his birthday. So we organized a minor celebration, presenting a cake to him at the start of the class. What I remember is thinking how really old he was, at all of forty three. But from the perspective of someone twenty plus years younger, that seemed very old indeed.

Opje went to Princeton as Director of the Institute for Advanced Study in the fall of 1947. At his invitation I followed along, transferring to Princeton University with the intention on both our parts that I would do a thesis with him. Once there, it became evident rather quickly that his days as a supervisor of Ph.D. theses were over. As the principal advisor to our government on atomic weapons and the future of atomic energy, he was away too much, and too preoccupied with his governmental responsibilities, to undertake thesis supervision.

But we still kept in close touch, as I would be invited to parties at Olden Manor, come to the Institute to hear great physicists lecture there, or see him at the colloquium that he, Rabi, and Wigner jointly organized. In June of 1948, he went back to Berkeley to give his famous course on quantum theory in their summer session. He knew that I was getting married just before the summer session began, and that I wanted to take part in his seminar as well. So he arranged for me to be his reader for the course.. Ed Gerjuoy has it quite right about that course. Namely, that after you listened to a lecture, you went home and spent hours writing it up, try to put in order the almost bewildering display of facts arising out of deep knowledge that was thrown your way. I had done that regularly during his lectures in 1946–47. Apparently he noticed this, because just before he gave his first lecture that summer in Berkeley in 1948, he said to me, "David, do you still have a copy of your notes on my lectures?" I said, "Yes," and that notebook containing my effort to transcribe his lectures was what he lectured from for the rest of the summer. It was a humbling experience but quite an inspiring one.

He was also an elegant and courtly individual. Shortly after we arrived that summer, I was walking to LeConte with Suzy, my brand new wife. We encountered Opje as he was returning from lunch. He looked at us, took off his hat, made a bow, kissed Suzy's hand and said to us, "This must be the new Mrs. Pines." As you can imagine, she was more than swept off her feet.

Later that summer, the two of us were invited to a family lunch. We started lunch with martinis, of course, at about 12:30 PM. We left about 5 PM. Of course I would love to be able to recall the details of the conversation but for the most part they are a blur. I remembered afterwards that this day was the third anniversary of the Trinity test, but not a word of that was spoken during the lunch.

Let me skip forward ten years to the spring of 1958, when the Oppenheimers made their first trip abroad after the security hearing and all of its terrible consequences for them. They arrived in Paris where he was to give a few lectures and be based at the École Normale. Opje was treated as a hero by the French. His appearance in Paris was quite comparable to that of a major rock star and he lectured to packed halls. *Paris Match* followed his every move. Cartier-Bresson appeared one morning at the École Normale to take photographs and a group of us were rounded up to be in the series he made of Oppenheimer in his office.

One day, Suzy and I took Opje and Kitty to a quite grand restaurant for lunch. As we were beginning to place our orders, the owner of the restaurant came over and asked if she could please have his autograph. A story they told us over lunch is one of our favorite Oppenheimer stories. A mutual friend of theirs knew Marlene Dietrich, and knew that she had written a little alphabet book in which she said, "O is for Oppenheimer, I wish I knew him." The friend arranged for them to meet and the evening came when Dietrich was to come to the Oppenheimers' apartment. About half an hour before she was due to arrive, Opje said to Kitty, "I think I should go downstairs just in case she gets lost." When he went downstairs, he found her pacing the street, waiting to come up. They had a wonderful time.

As a gesture of reconciliation on the part of our government and in recognition of Opje's extraordinary contributions during and after the War, Oppenheimer was awarded the Fermi prize of the Atomic Energy Commission in 1963. The presentation was to have been made by President Kennedy, but the timing was such that it took place just after his assassination, with Lyndon Johnson presenting the Award. What meant so much to Opje and Kitty was that Jackie Kennedy came and made a great point of saying how important the Award had been for Jack Kennedy, and how much he had looked forward to giving Opje the prize in person.

As Opje became ill and was treated for the lung cancer that killed him, his many, many friends would come for a farewell visit in Princerton. I was among those who saw him in January just before his death. He had words for each of us that related to his interaction with us through the years. It was not so much that we were helping him as he was helping us cope with the fact that he was dying.

His memorial service was a most moving event. Hans Bethe recalled his contributions to science; Harry Smythe his contributions at Los Alamos; and George Kennan talked about the larger role that Opje had played in the intellectual life of the country and of the world. Those quite extraordinary talks were then followed by a piece performed by the Julliard Quartet.

Let me turn briefly to Opje as a scientist. As graduate students at Berkeley, we would speculate on how it could have been that this absolutely brilliant man who was brighter than anyone we could imagine had never done work that would have brought him a Nobel Prize. We were simply ignorant of the fact he had in fact done so, in two seminal papers written in the late 1930s.

One was on the first serious calculation of the structure of a neutron star. With his student George Volkoff, he calculated the equation of state, radius, and likely maximum mass of a neutron star, a star made up almost entirely of neutrons, in which the gravitational attraction is balanced by the zero point energy the neutrons possess as a result of the Pauli exclusion principle. They found a radius of the order of seven kilometers, some four plus miles, while the average density of the neutrons was a little less than that of nuclear matter. These are amazing objects, containing as they do the most dense form of observable matter in the universe. They were only discovered some thirty years later, when pulsars were identified as rotating neutron stars.

With his student Hartland Snyder, he wrote the first paper on stellar gravitational collapse. Opje had realized that as the stellar mass increased, the gravitational attraction could not continue to be in balance with the Fermi pressure of the neutrons indefinitely. If you had too great a mass, the star would simply collapse, leaving behind a gravitational singularity. So, in a very real sense, he discovered both neutron stars and "black holes." If Opje had not been a chain smoker, if he had lived to a reasonable age, he would surely have received a Nobel Prize for this work sometime in the 1970s.

Let me conclude by saying a little bit about Oppenheimer the man. I have been lucky enough to know quite a few Nobel laureates and others whom you would classify as "geniuses." Opje stood out. He was probably the quickest of the group — the fastest to grasp the import of something new, the fastest to adapt to a changed circumstance. This brilliance was a total disaster for him growing up because it set him apart from all of his youthful counterparts, as many people have described. As a result, the barriers between him and his peers and later his colleagues were always there, because he thought much more quickly than almost anybody to whom he ever talked.

All great scientists are capable of holding more than one idea in their head at a time. Opje was capable of holding not two but maybe three of four. This, I suspect, is part of what led to his getting into problems during his wartime conversations with security people in which, as we heard quite convincingly today, he was trying to invent scenarios in order to protect his brother.

Opje leaned how to be a friend. It is clear that he probably had almost no friends until he was in his teens, probably into his twenties, but he developed a marvelous capacity for friendship. He became for many of us a caring and loving individual with whom it was a total pleasure to be around.

What further set him apart among people of remarkable creativity and intelligence was the capacity he developed for leadership. This was totally unexpected by almost anyone who knew him as a young man. As the leader of the only major school of American theoretical physics in the 1930s, he clearly had intimations of what he could do as a leader in physics. But this achievement was dwarfed by the extraordinary job he did during the three years he led the Manhattan Project.

He was as outstanding a leader as he was a scientist, mentor, teacher, and friend. All of us who knew him well still miss him and remember him with great affection and admiration.

Thank you.

J. ROBERT OPPENHEIMER: CONSUMMATE PHYSICIST

Maurice M. Shapiro
Professor, University of Maryland

Maurice Shapiro

In a memorial symposium celebrating the life and work of Robert Oppenheimer, it is natural that his role in hastening the end of World War II should be highlighted. It is this indispensable achievement that assures his place in history. As a wartime group leader under Oppenheimer in Los Alamos, I am acutely aware of the extraordinary qualities of leadership that enabled him to coordinate the efforts of many "prima-donnas." And yet there is another important facet of Oppie's career that should be remembered. He was a brilliant thinker and teacher who would have left an indelible imprint on science — and notably on physics in America — even if there had been no World War II. Since his major contributions as a theorist were to the science of quantum mechanics as applied to cosmic rays, I venture to emphasize these contributions in my talk. I am mindful that my audience is composed largely of non-physicists, so I should try to pride some background.

In 1939 Arthur Holly Compton convened an International Conference on Cosmic Rays, which attracted virtually all the leading physicists in the field. Among them were: Viktor Hess, the discoverer of cosmic rays; Carl Anderson, discoverer of the positron, who shared the Nobel Prize with Hess; Werner Heisenberg; Walther Bothe; Manuel Vallarta; Hans Bethe; Robert

Oppenheimer; Bruno Rossi; Pierre Auger; and Edward Teller. It was plain that Oppie was highly esteemed by his peers.

A School of Theoretical Physics

As a graduate student of Compton's, I was privileged to attend this conference at the University of Chicago, and that is how I first met Oppenheimer. Then 35 years old, he was already recognized as the founder of the first school of theoretical physics in the United States, which functioned on two campuses: the University of California at Berkeley and the California Institute of Technology. Twice each year, as Oppenheimer moved from one campus to the other, his graduate students followed him.

Nuclear Theory and Cosmic-Ray Showers

Ten years earlier, having mastered and contributed to the new science of quantum mechanics in Göttigen, Oppenheimer returned to the U.S. and joined the Berkeley faculty. Ernest Lawrence welcomed him as the "house theorist" who could make sense out of the findings in nuclear physics that were pouring out of experiments at the cyclotron. In Berkeley, Oppenheimer also taught quantum mechanics, and attracted a following of students and other collaborators.

Being impressed with the experimental program at Caltech, Oppenheimer accepted the concurrent offer of an appointment to teach quantum mechanics there as well. While engaged in low-energy nuclear theory, he realized that nature's accelerator — the cosmic radiation — offered the opportunity not then available at existing accelerators to probe more deeply into the nature of nuclear matter. The 1930's atmospheric showers of cosmic rays, revealed conspicuously in the work of Rossi and Auger, presented a challenge to theorists: could the multiplication of photons and electron pairs be elucidated by means of the existing theory of quantum electrodynamics? The effort to understand the observations of showers attracted the attention of leading physicists such as Bethe, Heitler, Bhabha and Nordheim. Oppenheimer and his associates, notably Carlson and Serber, were also active in this quest.

Cosmic-Ray Primaries

Compton's most important contribution to the field in the early thirties had been the demonstration, through a world survey of cosmic-ray intensities, of a variation with geomagnetic latitude. This showed that the "primary" cosmic rays incident upon the earth's atmosphere must be charged particles. The competing view, championed by Robert Millikan, asserted that the primaries are gamma rays — an assumption that had previously seemed reasonable in the light of the strong penetrating power of the "rays." Meanwhile, Rossi suggested that a comparison of intensities from the east and west would reveal whether the primaries were mainly positively charged or negative, i.e., as generally supposed, positrons or electrons. East–West experiments showed that the primaries were mainly (if not entirely) positive. Not until 1941, in a memorable balloon investigation by Schein, Jesse and Wollan, was it demonstrated that the primaries had to be mainly protons.

In a brilliant cosmic-ray review (published the same year, but evidently written in 1940) Oppenheimer was unaware of the Schein experiment, and so he supposed that the puzzle of the cosmic ray primaries was still unsolved. It is worth noting that three decades had elapsed after Hess' balloon flights before the identity of the main component of the primaries was revealed. And another six years elapsed before it became clear that heavier nuclei than those of hydrogen also arrived among the primaries. In due course it was shown that virtually the whole periodic table of the elements — and many isotopes of those elements — comprised the rich array of the cosmic-ray primaries. Schein's important contribution remained almost unknown for some years because it was published when most physicists were too busy with wartime research to read *The Physical Review*. It is likely that Oppenheimer, fully engaged in helping to plan and organize the Manhattan Project, was among those who missed this discovery.

Showers: Radiation and Pair-Production

We return to the problem of electromagnetic cascades to which Oppenheimer and his collaborators contributed significantly in the 1930s. These showers of electrons and photons were called the "soft" component, as they were readily absorbed in dense materials. They are propagated by a succession of radioactive and pair-production events. When an energetic electron or positron collides with an atom, a photon (X-ray or gamma-ray) is produced. When, in turn, the energetic photon strikes an atom, it can produce an electron–positron pair. The repetition and propagation of these radioactive and pair-producing processes results in a multiplication of particles, generating a shower. When the energy of the initiating particle exceeds some million billion electron-volts, the resulting cascade is known as an EAS (extensive air shower), first discovered by P. Auger.

Mesotrons (Mesons)

In addition to the soft component, a "hard" component was observed, which could penetrate great thicknesses of iron or lead. These turned out to be "mu mesons," intermediate in mass between electrons and protons. P. Blackett and G. Occhialini, among others, used cloud chambers in magnetic fields to study the nature of shower particles. Their photographs revealed tracks of particles that exhibited greater magnetic rigidity than those of electrons. These were evidently due to the penetrating particles observed by Anderson and others. They came to be called "mesotrons" (subsequently "muons"), and seemed to possess some properties of the short-lived "nuclear-glue" particles postulated by H. Yukawa. These muons had a mass in the range of 100–200 electrons masses, and a lifetime of about 2 microseconds.

Could the Muons be Yukawa Particles?

The muons presented a tantalizing puzzle. Despite their similarity in mass and lifetime to Yukawa's particles, their great penetrating power contradicted an essential attribute of mesotrons, i.e., their strong interaction with nuclear matter. In due course, the puzzle was solved, first by theorists, and later by experimental confirmation. R. Marshak and H. Bethe proposed a two-meson hypothesis: a parent meson and its radioactive daughter. The first, assumed to be produced in violent collisions near the top of the atmosphere, was a very short-lived Yukawa-type particle. Its charged decay product was the penetrating particle detected in cloud chambers.

Soon, in 1946, this hypothesis was verified by Powell, Occhialini, and Lattes, who had exposed sensitive photographic emulsions at a mountain altitude. They found microscopic tracks of what they called pi mesons (or pions) decaying into mu mesons (muons). The muons decayed, in turn, giving rise to electrons. The mystery was solved: the anticipated Yukawa particles were the pions, with a lifetime of roughly one percent of a microsecond. In this time interval they could hardly travel very far before their disintegration into the longer-lived, penetrating muons. The latter could survive down to sea level and even penetrate deep underground, thanks to time-dilation of relativistic particles. Evidently the muons constituted the hard component observed at sea level. Soon it was realized that neutrinos were also emitted among the decay products of both pions and muons.

Neutral Pions and Showers

Although cosmic-ray showers were extensively studied, their origin in the upper atmosphere was a mystery. According to B. Rossi, Oppenheimer proposed in 1947 the existence of neutral pions that decayed promptly into pairs of gamma rays, thereby initiating electromagnetic cascades. This hypothesis was soon verified experimentally. The discovery of charged pions which gave rise to the hard component (via their daughter muons) could hardly have been anticipated in the 1930s, however, the elucidation of cascade development in the atmosphere demonstrated the power of quantum mechanics to account quantitatively for the successive (alternating) processes of radiation and pair production that generated cosmic-ray showers. In this important application of electromagnetic theory, Oppenheimer and his co-workers played a notable part between 1930 and 1937. His interest in cosmic rays continued until 1941, and was resumed after the interruption due to World War II.

Of some eighty of Oppenheimer's published papers in physics, 29 were devoted to cosmic-ray theory. In some of these he enjoyed the collaboration of students and colleagues (see Appendix below). His other principal contributions were in quantum mechanics, nuclear theory, and the theory of neutron stars.

If this talk has emphasized cosmic-ray theory, this may be understood in view of Oppenheimer's great interest in and significant contributions to the field. Also, I confess that, since my own career in physics has been devoted mainly to working in this discipline, I felt comfortable in discussing it.

In conclusion, I thank Ms. Cindy Kelly for inviting me to participate in this symposium in fond memory of a great scientist and a good friend.

Appendix: Oppenheimer's Principal Collaborators in Cosmic-Ray Theory

H.E. Bethe, J.F. Carlson, R.F. Christy, W.H. Furry, H.W. Lewis, L. Nedelsky, E.C. Nelson, L.W. Nordheim, J. Schwinger, R. Serber, S.A. Wouthuysen.

A FEW WORDS FROM AN OPPENHEIMER

Andrew R. Oppenheimer

It is truly a great honor for me to be a part of this very special celebration this weekend. I am very grateful to Cindy Kelly for asking me to speak at the dedication of the house where Robert Oppenheimer and his family lived here in Los Alamos during the Manhattan Project, and at this Symposium. Thank you for inviting me.

Andrew R. Oppenheimer
Photo by Rick Scibelli

I am proud beyond measure to talk about a man who has influenced the world more than most people both in the U.S. and beyond realize today, and about the effect he has had on my life. Indeed, it is my fascination with the life of Robert Oppenheimer, the Manhattan Project, and the Cold War that brought me to my eventual career — as a nuclear weapons expert for Jane's Information Group and other institutes. In this role I am immersed daily in the technology and issues surrounding weapons of mass destruction. I feel a heavy responsibility in these troubled times we live in. In my own small way I am trying to continue the Oppenheimer tradition: to advise and communicate on weapons issues and problems.

Since childhood I knew that this great and famous man whose name I am so proud to bear had changed the world forever. As a student of physics, and of politics, in Liverpool where I grew up, I became fascinated by the drama and tragedy of the Bomb. I devoured everything I could find about nuclear weapons — both the science and the politics — and everything about Robert Oppenheimer. His science, his Bomb, his terrible trial, his life.

"Father of the Atomic Bomb." This nomenclature may be a cliché but it is a title that echoes down the years, as it represents in a few words what Oppenheimer's role was in ushering in the world's first, true, weapon of mass destruction, in the midst of the world's worst-ever war. Those great and good scientists who made the atomic bomb under his leadership believed that the Nazis were developing one — the very essence of the unthinkable. They had every right to believe that. When this didn't happen, events had over-taken them.

I first came to Los Alamos in the late 1980s. I got to know the town and the many lovely people who have made me welcome during seven visits. I soaked up this place's extraordinary history and tried to imagine how it was when it all began. So, as the Centennial approached, I hoped so much that I could come over to join you in celebrating one of the greatest Americans of the 20th century — an icon of science and of leadership, of creation and destruction, and above all, of great humanity. There is no doubt that he was the most important nuclear scientist in history, who was not rewarded enough in his life.

In the early 1980s I saw the monumental BBC seven-part drama about Robert which won drama awards in Britain. I spoke at my first conferences about nuclear weapons — particularly on missile defense, which was an emerg-ing policy. At that time — at the height of the second Cold War — people were increasingly taking an interest in the nuclear arms race and the history of the Bomb. There now needs to be a revival about the subject on TV and also in the movies — but any film about Robert will have to be of really great quality if it is to equal that BBC series.

Things have changed since the Cold War crisis that we were going through then, back in the 1980s. Today, in the post-9/11 world, the emphasis is on proliferation of weapons of mass destruction and the growing possibility of terrorists using nuclear or radiological weapons. These are subjects I special-ize in and about which I speak at conferences. I am interviewed by the media about these problems and about events as they unfold. It is a job I do with enormous pride and which is full of resonances for me.

The nuclear terrorism issue has a special resonance. As early as the late 1940s, Robert Oppenheimer was asked at a congressional hearing how author-ities would detect a nuclear weapon in an incoming shipping crate. The answer was: "with a screwdriver" — meaning you would have to open up every crate that came into American ports. The Atomic Energy Commission then commissioned a panel to study how to detect and prevent nuclear weapons from being smuggled into the country. It became known as the Screwdriver Report and remains classified to this day — never was it more relevant than now.

Our world continues to search for heroes. We must therefore rehabilitate this hero of the 20th century in the eyes of America and the world — a world that still lives with nuclear weapons. Robert Oppenheimer's actions, however flawed they were judged to be, expose the eternal dilemma of the scientist in our modern age. This dilemma is just as great today, when we face a revolution in biotechnology, enabling us with the potential to construct incurable diseases that could wipe out millions. And to evolve other technologies with similar portent, such as could enable a fourth generation of nuclear weapons and exotic explosives with the equivalent power of a hydrogen bomb.

As I deal every day with these issues I feel Robert's spirit guiding me. If I come upon a problem, or have to speak on these issues, or encounter a difficult challenge, I feel Robert's spirit guiding me. I only hope to God that I have attained just one atom of his brilliance and wisdom.

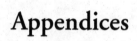

Appendices

APPENDIX I — AGENDA FOR OPPENHEIMER AND THE MANHATTAN PROJECT

"Oppenheimer and the Manhattan Project" Los Alamos, NM
Friday, June 25 and Saturday, June 26, 2004
AGENDA 6/17/04

Friday, June 25, 2004

10:00 AM; 12:00 PM; 1:30 PM and 3:00 PM Tours of the Manhattan Project at Los Alamos

> Tours will begin at the Bradbury Science Museum and will bring to life the Los Alamos of the Manhattan Project from the technical areas to the top-secret community. Historians, Manhattan Project veterans and family members will augment the professional guides for a lively and informative narrative.

9:00 AM – 4:00 PM Special Programs at the Bradbury Science Museum

> Visitors will enjoy an overview of the history of Los Alamos and a virtual tour of the Manhattan Project properties that are now located "behind the fence" with free slide show offered by the Los Alamos National Laboratory at 11 AM, 1 PM and 3 PM.

> An 18-minute documentary film, "The Town that Never Was," will be shown at 10 AM, 12 PM and 2 PM with reminiscences from Manhattan Project veterans John Mench (@10:30), McAllister Hull (@12:30) and Paris Howard (@2:30).

1:00 PM – 4:30 PM Book Signing by Authors at Fuller Lodge
Richard Rhodes, author of *The Making of the Atomic Bomb*, will be available for book signing with Jon Hunner and Robert Norris from 1:00 PM to 2:45 PM and with Gregg Herken, Joseph Kanon and Ferenc Szasz from 3:00 PM to 4:30 PM. The Los Alamos Historical Society will have copies of the books available for sale.

4:00 PM – 5:00 PM Open House — J. Robert Oppenheimer's Home
Helene and Gerry Suydam have graciously opened their home so that the public can see where the Oppenheimer family lived during the Manhattan Project.

5:00 PM – 6:00 PM Dedication of the Oppenheimer House at the Rose Garden behind Fuller Lodge, Los Alamos

U.S. Senator Pete Domenici, Governor Bill Richardson, LANL Director Pete Nanos, Chairman Nona Bowman, Los Alamos County Council, Nancy Bartlit, Los Alamos Historical Society, Ernest Ortega, National Park Service, Santa Fe, NM; Peter Wirth, NM State Representative, and Andy Oppenheimer, cousin of J. Robert Oppenheimer.

6:00 PM – 9:00 PM Reception and Dinner at Fuller Lodge
Brief remarks by LANL Director Pete Nanos and others

Saturday, June 26, 2004: Symposium on Oppenheimer and the Manhattan Project

8:00 AM – 9:00 AM	Registration at the Smith Civic Auditorium, Los Alamos
9:00 AM	Welcoming Remarks — Cynthia Kelly, President, Atomic Heritage Foundation; Chairman of Los Alamos County Council Nona Bowman; Director Pete Nanos
	The Significance of the Manhattan Project: A National Perspective Senator Jeff Bingaman, U.S. Senator for New Mexico

Standing on the Shoulders of Giants
Deputy Administrator Everett Beckner, National
Nuclear Security Administration

The Manhattan Project and New Mexican History
Stuart Ashman, Director, Office of Cultural Affairs,
State of New Mexico

10:00 AM   *Oppenheimer: King of the Hill*
Richard Rhodes — Keynote Address

10:45 AM   Press Opportunity (Break)

11:15 AM   *Oppenheimer — Years Before the Manhattan Project*
Jon Hunner, "The Early Years of Robert
Oppenheimer"

11:35 AM   *Oppenheimer and the Manhattan Project*
Robert Norris, "General Groves' Indispensable
Scientist"

Kai Bird, "Oppenheimer: The Manhattan Project
Years"

12:15 PM   Lunch Break

1:15 PM   *Alternative Perspectives*
Gregg Herken, "The Cautionary Tale of Robert
Oppenheimer"

Joseph Kanon, "A Novel Idea of Oppenheimer"

2:10 PM   *Reflections on Oppenheimer*
Ferenc Szasz, "Oppenheimer and New Mexico"
Maurice Shapiro, "J. Robert Oppenheimer —
Consummate Physicist"

3:00 PM   Break

3:20 PM   *Oppenheimer Remembered: Personal Vignettes*
Ed Gerjuoy, "Oppenheimer as a Teacher of Physics
and Ph.D. Advisor"

Andy Oppenheimer, "A Few Words from the Oppenheimer Family"

David Pines, "Remembering Opje: Teacher, Scientist, and Friend"

4:20 PM   *Preserving the History of the Manhattan Project*
Cynthia Kelly, "Defying the Odds"

5:00 PM – 6:00 PM   Book Signing

APPENDIX II — CONTRIBUTORS

Stuart Ashman is the Cabinet Secretary for the Department of Cultural Affairs for the State of New Mexico, where he is responsible for the oversight and vision for a complex group of cultural institutions, including the New Mexico Historic Preservation Division. He has served as the director for several museums, including the Governor's Gallery and the Museum of Fine Arts in the Museum of New Mexico as well as most recently at the Museum of Spanish Colonial Art.

Everet Beckner is Deputy Administrator for Defense Programs at the Department of Energy's National Nuclear Security Administration (NNSA), responsible for the nation's nuclear weapons complex. He recently retired as Vice President at Lockheed Martin and previously served as the Energy Department's Principal Deputy Assistant Secretary for Defense Programs (1991–1995). He has also worked at the Sandia National Laboratories and has a Ph.D. in physics.

Senator Jeff Bingaman graduated from Harvard University in 1965 and earned a law degree from Stanford University in 1968. After a year as New Mexico Assistant Attorney General and nine years in private law practice, he was elected Attorney General of New Mexico in 1978 and to the U.S. Senate in 1982.

Kai Bird is a Fellow at the Woodrow Wilson International Center for Scholars. He is the biographer of John J. McCloy and McGeorge and William Bundy and co-editor of *Hiroshima's Shadow: Writings on the Denial of History and the Smithsonian Controversy*. He is co-author with Martin Sherwin of *American Prometheus: The Triumph and Tragedy of J. Robert Oppenheimer*.

Nona Bowman was elected chair of the County Council in January 2004. She brings a record of community service to this position extending from

Los Alamos back to Gaithersburg, Maryland and to Livermore, California. Her first action upon arriving in Los Alamos was co-chairing the first renovation of Fuller Lodge in 1986, which arrested the deterioration of the Lodge and helped establish it as the centerpiece site for Los Alamos history.

Senator Pete Domenici graduated from the University of New Mexico in 1954 and earned a law degree from the University of Denver in 1958. Domenici spent the time in between his two degrees teaching math at Garfield Junior High and pitching for the Albuquerque Dukes, a farm team of the Brooklyn Dodgers. He was elected to the Albuquerque City Commission in 1966 and then the Senate in 1972. With re-election in 2002, Domenici became the first New Mexican elected to serve six terms in the Senate.

Edward Gerjuoy is Professor of Physics Emeritus at the University of Pittsburgh. In 1977, after a full career in physics teaching and research, he obtained a J.D. degree; thereafter, he has divided his time between physics and environmental law. He is the author of more than 100 physics publications, 40 papers on legal and public policy issues, and about 100 adjudicating opinions while serving as a member of the Pennsylvania Environmental Hearing Board.

Gregg Herken is an historian and professor in the School of Social Sciences, Humanities and Arts at the University of California, Merced. Formerly, he was the Curator of Military Space History at the Smithsonian Institution's National Air and Space Museum in Washington, DC. He is the author of three books on nuclear history and a biography, *Brotherhood of the Bomb: The Tangled Lives and Loyalties of Robert Oppenheimer, Ernest Lawrence, and Edward Teller* (Henry Holt and Company).

Jon Hunner directs the Public History Program at New Mexico State University where he also teaches U.S. history. His first book, *Inventing Los Alamos: The Growth of an Atomic Community*, was released in the fall of 2004. His next book, tentatively titled, *Chasing Oppie: J. Robert Oppenheimer and the American West*, is scheduled for release in 2007.

Joseph Kanon has written four novels, *Los Alamos, The Prodigal Spy, The Good German* and *Alibi*. *Los Alamos,* a historical thriller about the Manhattan Project in the spring of 1945, was an international bestseller, translated into fifteen languages, and won the Edgar Award.

Cynthia C. Kelly is President of the Atomic Heritage Foundation, dedicated to the preservation of the history of the Manhattan Project and the Atomic Age. For over twenty years, she was a senior manager at the Environmental Protection Agency and Department of Energy and received a Distinguished Career Service Award in 1999.

Pete Nanos is the Director of Los Alamos National Laboratory. During his distinguished naval career, he commanded the strategic nuclear program and served as Commander, Naval Sea Systems Command for the Navy. Upon leaving the Navy, he served as principal deputy associate director for Los Alamos National Laboratory's Threat Reduction Directorate before taking the Director's position with the Los Alamos National Laboratory.

Robert S. Norris has been a research associate for almost twenty years at the Natural Resources Defense Council in Washington, DC, covering nuclear weapons issues. As an author of the multi-volume Nuclear Weapons Databook series, and of numerous articles, he has written extensively about the nuclear programs of the United States, Soviet Union/Russia, Britain, France, and China. He is the author of *Racing for the Bomb: General Leslie R. Groves, the Manhattan Project's Indispensable Man* (Steerforth Press, 2002).

Andrew R. Oppenheimer, cousin of J. Robert Oppenheimer, is a nuclear weapons expert and consultant on weapons of mass destruction for governments and institutes in Britain, where he is based, and the United States. He has degrees from the University of Liverpool, King's College London and the Open University. Andy Oppenheimer is also an artist whose portraits of J. Robert Oppenheimer were shown in Los Alamos in 2003.

David Pines is a theoretical physicist who is the founding Co-Director of the Institute for Complex Adaptive Matter and Research Professor of Physics and Professor Emeritus of the Center for Advanced Study, University of Illinois at Urbana-Champaign. His current research is focused on emergent behavior in unconventional superconductors. His contributions to the theory of many-body systems and to theoretical astrophysics have been recognized by two Guggenheim Fellowships, the Feenberg Medal, Friemann, Dirac, and Drucker Prizes, and by his election to the National Academy of Sciences, American Philosophical Society, the Russian and Hungarian Academies of Sciences, and other scholarly societies.

Richard Rhodes is the author of 20 books, including *The Making of the Atomic Bomb*, which won a Pulitzer Prize in Nonfiction, and *Dark Sun*, one of three finalists for a Pulitzer Prize in History, that continued the story of nuclear weapons development in the early Cold War years. Rhodes has written extensively about nuclear issues and lectured widely in the United States and abroad.

Maurice M. Shapiro is Visiting Professor, University of Maryland. After the Manhattan Project, he had a distinguished career in the field of cosmic rays and neutrino astrophysics. Dr. Shapiro is director of the International School of Cosmic Ray Astrophysics that holds biennial courses for graduate students and young researchers in Erice, Italy.

Ferenc Szasz is Regents' Professor of History at the University of New Mexico where he has taught American Social and Intellectual History for over three decades. He has published several articles and two books on the early atomic world: *British Scientists and the Manhattan Project: The Los Alamos Years* and *The Day the Sun Rose Twice: The Story of the Trinity Site Nuclear Explosion, July 16, 1945.*

Index